THE LIGHT COURSE

[XXII]

FOUNDATIONS OF WALDORF EDUCATION

RUDOLF STEINER

The Light Course

FIRST COURSE IN NATURAL SCIENCE:
LIGHT, COLOR, SOUND—
MASS, ELECTRICITY, MAGNETISM

TRANSLATED BY RAOUL CANSINO

 Anthroposophic Press

Published by Anthroposophic Press
P.O. Box 799
Great Barrington, MA 01230
www.anthropress.org

Translation copyright © 2001 by Anthroposophic Press

This work is a translation of *Geisteswissentschaftliche Impulse zur Entwickelung der Physik: Erster naturwissenschaftlicher Kurs: Licht, Farbe, Ton—Masse, Elektrizität, Magnetismus* (GA 320); copyright © 1964 Verlag der Rudolf Steiner–Nachlassverwaltung, Dornach, Switzerland. Translated with permission.

Publication of this work was made possible by a grant from the Waldorf Curriculum Fund.

Book design by Jennie Reins Stanton.

6/03

Library of Congress Cataloging-in-Publication Data

Steiner, Rudolf, 1861-1925.
[Lichtkurs. English]
The light course : ten lectures on physics : delivered in Stuttgart, December 23, 1919-January 3, 1920 / by Rudolf Steiner ; translated with a foreword by Raoul Cansino.
      p. cm. -- (Foundations of Waldorf education ; 22)
      ISBN 0-88010-499-6
   1. Light. 2. Color. 3. Anthroposophy. I. Title. II. Series.
      QC361 .S8313 2001
      535--dc21

                                                              2001003239

10 9 8 7 6 5 4 3 2 1

Printed in the United States of America

Contents

Translator's Introduction

On a parent education evening at Green Meadow Waldorf School in New York, the class teacher of the seventh grade demonstrates a first physics experiment for the parents in attendance. Over a Bunsen burner he heats a beaker of water containing a piece of ice. The parents watch in rapt silence for several minutes while tiny bubbles form on the bottom and sides of the beaker. Losing its milky opacity and gradually taking on the transparency of the surrounding water, the chunk of ice becomes more mobile, swimming about slowly in the beaker. Bubbles begin to form around the piece of ice, and, one by one, little bubbles rise from the bottom of the beaker, describing erratic paths to the surface. Soon the chunk of ice is no more than a ghostly semblance of its former self, perceptible only as a fleeting watery "thickness" or as a sensation of movement. Then, with surprising suddenness, the water itself is full of motion and no longer transparent but turbulent with large bubbles that swiftly ascend the sides of the beaker. The water itself appears to flow upward and then toward the center of the surface, where it seems to be sucked down again into the boiling cauldron. Surprisingly, very little steam is generated in this process, but when the teacher turns off the Bunsen burner, steam suddenly becomes visible, rising from the now quiet water, in which there is no more ice to be seen. The ice has "melted." The parents then offer their observations. What did they see?

For many of the parents, it is a first glimpse into the phenomenally based science curriculum that their children have been learning since their botany block in fifth grade. For the class teacher, it is an opportunity to explain that Waldorf education aims to bring the children an understanding of the physical world that is based on what they can actually observe with their senses. After observing such an experiment, the children attempt to put into their own words what they have seen. If they say that the water boiled and the ice melted, the teacher encourages them to describe the actual individual moments until the class has built up a full picture of the process. The children are learning (or actually relearning) how to attend to a natural phenomenon without substituting concepts such as "boil" or "melt" for actual perceptions. This sense-based way of doing science, which has its roots in Goethe's scientific practices, is to continue throughout the children's education even through the high school.

As a dyed-in-the-wool friend of the humanities, who as a schoolboy had avoided the "hard" sciences whenever possible, I was fascinated by both the demonstration and the explanation. As a student of German literature, I had heard about Goethe's ideas on color and had a passing acquaintance with the controversies surrounding the great poet's work in science. A subsequent Waldorf conference, at which science teachers Stephen Edelglass and Michael D'Aleo spoke about the Goethean approach to physics, once again piqued my interest: here was a way of looking at the natural world without reducing it to dry formulas and invisible forces. Where had this approach come from?

"We can definitely stick with the phenomenon. That is good," said Rudolf Steiner in the "Discussion Statement" (August 8, 1921) that has been printed here in lieu of an afterword to *The Light Course*. A simpler description of

Goethe's approach could hardly be given, yet it captures the essence: Goethe was not interested in "natural laws," in finding a cause lurking behind the phenomena. Instead he sought by dint of careful observation to create what Steiner called "a kind of rational description of nature" (First Lecture), which would reveal the "archetypal phenomenon" (*Urphänomen*), consisting of the most basic elements of the observed phenomena. Goethe saw such an archetypal phenomenon in the colors that appeared when he first looked through a spectrum toward a window where the darkness of the frame met the brightness of the sky.

"First Course in Natural Science" was the name Rudolf Steiner originally gave to this series of ten lectures for the teachers of the new Waldorf School in Stuttgart from December 23, 1919, to January 3, 1920. Over the intervening years these lectures gained the sobriquet "The Light Course," a misnomer perhaps, since the course deals with a much larger range of phenomena, encompassing, besides light and color, discussions of sound, mass, electricity, and magnetism, and even venturing into areas such as radioactivity, relativity, and quantum mechanics, which constituted the cutting edge of physics at that time. Nevertheless the nickname does have a certain justification, since all of lectures three through seven and a good deal of lecture two are devoted to light and the related phenomenon of color. Equally significant, the discussion of light gave Rudolf Steiner the opportunity to establish the phenomenological approach of Goethe's *Color Theory* as the methodological basis for looking at other physical phenomena. Far from being a straightforward guide to teaching physics in the Waldorf School with practical suggestions on curriculum and teaching methods, *The Light Course* and two subsequent courses on the natural sciences given in 1920 and 1921 were intended as a basic schooling in the Goethean approach to science and as an introduction to Rudolf Steiner's project of anchoring natural science in a science of the spirit.

At its core *The Light Course* is a critique of the materialistic thinking of modern science that separates the perceived object from the perceiving subject, denying the inner spiritual experience of the human being and reducing consciousness to a mere artifact of stimulated matter. Steiner poses the basic epistemological question: how do we know what we know? He contrasts the purely abstract "mathematical way of looking at natural phenomena" characteristic of classical science with an approach based on human beings and their relationship, through the senses, to the natural world. By reclaiming the validity of sensory experience, Steiner bridges the chasm between the inner experience of the human being and the "real" outer world. Guiding his audience through a series of classic physics experiments, Steiner interweaves an intensely sense-based treatment of the phenomena with the insights of spiritual science, anthroposophy, coming to conclusions that are of interest to scientists, teachers, and students of philosophy alike.

The Light Course was given little more than a year after the armistice that ended World War I, a war in which modern technology had powerfully magnified the forces of destruction. In the aftermath of the horrors inflicted on humanity in this war, Steiner was deeply concerned about the use—and abuse—of scientific knowledge. In their book on Goethean science, *The Marriage of Sense and Thought*, Stephen Edelglass, Georg Maier, and their coauthors remark that there is a moral dimension to the study of nature:

> Human beings are creating a world that is increasingly inhospitable to themselves or anything else alive. The empathetic basis on which we relate to nature is eroded, as is that on which we relate to each other and

to our own selves. Our impotence to reverse these trends derives from our unquestioning acceptance of the hypothetical-reductive-mathematical methods of science. We seem to feel that such methods are logically necessary. Reductionists are convinced that objective knowledge can be gained by no other means. However, built into these methods is the unsupported presupposition of a reality that, in its finality, is static, fragmented, and impersonal. Within such a reality there is no place for life or sentient human beings.[1]

Steiner warns of this danger in the concluding words of the last lecture of *The Light Course*, when he refers to the collaboration that took place during the First World War between the military and the physics departments of the universities:

My dear friends, the human race must change its ideas, and it must change them in many areas. If we can decide to change them in such an area as physics, it will be easier for us to change our ideas in other areas too. Those physicists who go on thinking in the old way, however, won't ever be far removed from the nice little coalition between the institutes of experimental science and the general staffs.

In *The Light Course* Steiner proposes phenomenological science as a path to change the consciousness of humankind, a path that leads away from the fragmentation and alienation of modern culture toward a new understanding of the place of the human being in the wholeness of nature. Steiner's desire to help us find this path was the impulse that led to the founding of the first Waldorf school. When the children in a Waldorf school study the natural sciences, from their introduction to botany in the fifth grade to their investigations of optics in the twelfth, they themselves, with their physical experience of the world and their thoughts about these experiences, are at the center of the study. Thus when the bubbles begin to form around the ice in the beaker of water, the Waldorf teacher's first concern is not that the children should "know" the boiling and freezing points of water, but that the children's sense experience should lead to an inner understanding of nature—a kind of "knowing" that doesn't rely on theory alone, but on the children's sense of their place in the natural world—bridging the chasm between the water bubbling in the beaker and the thoughts bubbling in the child's mind.

Raoul Cansino
Chestnut Ridge, New York, 2001

A Note on the Text

Rudolf Steiner's lectures were influenced by the social life in the circle of his students and by their needs and the demands of the moment. Many of the lectures are answers to questions that were living in the circle of the listeners. Repeatedly the situation is that of a response to questions, of a conversation. We owe these lectures on physics to this extemporaneous speaking, which, despite its immersion in the context of the moment, is always directed toward larger developmental perspectives. The immediate occasion for the lectures was an inquiry from the faculty of the Waldorf School, which had been founded only a few months earlier under the direction of Steiner. The participants in the course were, for the most part, the teachers of the Waldorf School. Thus what came about within the smallest of circles reaches far beyond this circle in its essence.

Parallel to this course, Steiner also became intensively active in various other directions, for the development of the Waldorf School and, in general, for the transformation of social relations in a spiritual sense: conferences with the teachers, a course they had requested on "Linguistic Observations of Spiritual Science," social science lectures for the public, lectures to the members of the Anthroposophical Society, conferences and discussions for the enterprise "Der kommende Tag" ("The Coming Day"). All of this made the 1919 Stuttgart Christmas season one of the richest creativity but also one of great demands.

In keeping with their genesis, these lectures were not intended for print. Accordingly, the transcription and drawings were not corrected by the lecturer. It is only to be expected that the rendering is not always faithful to the original meaning. If this can be said of the majority of Steiner's lectures, it is particularly true for these physics lectures, in view of the difficulties that attend the transcription of experimental presentations of this kind.

Printed in lieu of an afterword to the course is a statement from a discussion that serves to clarify the meaning and character of these physics presentations in a concise way.

Text documentation: An official stenographer was not engaged for the course. The text of the typewritten version was worked up on the basis of the shorthand record of various participants, according to a note from Helene Finckh, the official stenographer in Dornach and for most of the other lectures, starting in 1916. No other details are known about how the text was produced. The German edition that this translation is based on followed this text very closely. The notes are those of the editors of the German edition unless otherwise noted.

The editors of the Rudolf Steiner Verlag gave the volume the title *Geisteswissenschaftliche Impulse zur Entwickelung der Physik* ("Impulses from Spiritual Science for the Development of Physics"). Originally, it was called *Erster naturwissenschaftlicher Kurs* ("First Course in Natural Science").

First Lecture

STUTTGART, DECEMBER 23, 1919

FOLLOWING UP ON the words just read to us here,[1] some of which are already over thirty years old, I would like to remark that, in this brief time at our disposal, I will only be able to provide you with highlights about the study of nature. First of all, especially since we do not have very much time, we can continue what we have begun here in the near future;[2] and, second, since I was informed of the intention of having such a course only after I arrived here, for the time being it will be a very episodic matter indeed.

On the one hand, I want to give you something that can be usable for the teacher, perhaps less in the sense that it can be used directly as lesson content than in the sense that it can inform your teaching as a certain basic scientific direction. On the other hand, given the multiplicity of contradictory theories presently circulating, especially in the natural sciences, it is particularly important for the teacher to have the right idea as a basis. With this in mind, I would also like to give you a few pointers.

I would like to add something to the words that Dr. Stein has just so graciously recalled—something that I found myself forced to say at the beginning of the 1890s, when I was invited by the Frankfurt Free Seminary to give a lecture on Goethe's natural science.[3] In my opening remarks at that time I said I would have to limit myself to speaking primarily about Goethe's relationship to the organic sciences, since injecting the Goethean worldview into the study of physics and chemistry

was a sheer impossibility. It is impossible simply because physicists and chemists are condemned by everything that presently
exists in physics and chemistry to regard everything coming
from Goethe as a kind of nonsense, as something that is mean-
ingless to them. At that time I expressed the opinion that we
would have to wait until physics and chemistry were led by
their own research, so to speak, to realize that the structure of
their scientific effort was leading to absurdity. Only then would
the time come when Goethean views could also take root in the
fields of physics and chemistry.

Now I will try to reconcile what we might call experimental natural science with what we gain by the results of experimentation. I want to say a few words by way of introduction
and theoretical explanation. Today I am aiming to work toward
a real understanding of the distinction between popular, every-
day natural science and the scientific ideas that can be derived
from Goethe's general worldview. First, however, we will have
to go a bit into the theoretical premises of scientific thinking.
Those who think about nature today in the popular sense usually have no clear idea of what their real field of research is.
Nature has become a vague concept. Therefore we do not want
to begin with the popular view of the essence of nature, but
rather with the way we normally work in the natural sciences.
This way of working, as I am going to characterize it, is in fact
somewhat caught up in transformation, and there is much we
could interpret as the dawn of a new worldview. But, on the
whole, the way of working that I am going to characterize for
you today still predominates.

Today researchers try to approach nature from three starting points. First, they try to observe nature in such a way that
on the basis of natural beings and phenomena they arrive at
concepts of species and genera. They try to classify natural phenomena and beings. You need only recall how these appear to

people in outward sense experience, for example, individual wolves, individual hyenas, individual heat and electrical phenomena, and how researchers try to combine such individual phenomena and group them in species and genera, speaking of the species wolf, the species hyena, etc., and also of certain categories of natural phenomena—in other words, how they group things that exist individually. We might say, however, that this activity, though important, in natural science is actually practiced in a somewhat underhanded way. We are not aware that we would actually have to investigate how the general category we have arrived at by dividing and classifying is related to the individual phenomenon.

The second thing we do these days when we are active in the field of natural science is to try to find what we call the causes of the phenomena, either by preliminary experimentation or by the following step, the conceptual processing of the experimental results. When we speak of causes, we often have forces or materials in mind: we speak of the electrical force, the magnetic force, heat, etc. But often we have something more comprehensive in mind. Behind the phenomena of light or electricity we speak of an unknown such as the ether. We try to derive the characteristics of this ether from the results of experiments. You are aware that everything said about this ether is extraordinarily controversial. However, one thing can certainly be pointed out: in the attempt to arrive at the causes of phenomena, we are seeking the way from the known to an unknown, although without inquiring much about the justification for proceeding from the known to the unknown. For example, when we perceive some light or color phenomenon, which we describe subjectively as a color quality, we hardly take into account what right we have to speak as if the effect on us, on our soul, on our nervous system, were the effect of an objective process that takes place as a wave movement in the cosmic ether. Thus we would actually have to distinguish two things: the subjective process, on the one hand, and the objective process, which consists of a wave movement of the ether or of the interaction of the latter with the processes in perceptible matter.

This way of looking at things—which is beginning to become a bit shaky—is the one that dominated the nineteenth century and, in fact, is still ubiquitous in the way we speak of phenomena, continuing to permeate our scientific literature; it permeates the way we speak about things.

Then there is the third way by which so-called natural scientists attempt to approach the configuration of nature—by looking at the phenomena. Let's take a simple phenomenon. If we drop a stone, it will fall to the earth, or if we tie it to a string and let it hang, it will pull in a vertical direction toward the earth. We collect such phenomena and arrive at what we call a natural law. Thus we regard it as a simple natural law when we say that every planetary body attracts the bodies located on it. We call this force gravity and explicate it in certain laws. The three laws of Kepler, for example, are a paradigm for such laws.

So-called natural science attempts to approach nature in these three ways. Now I want to contrast how the Goethean view of nature actually strives to do the opposite of all three. First of all, when Goethe began to occupy himself with natural phenomena, he found the classification of natural beings and facts into species and genera highly problematic. He questioned the validity of inducing certain rigid concepts of species and genus from individual concrete beings and concrete facts. Instead he wanted to pursue the gradual transformation of one phenomenon into another, to follow the transformation of one state of a being into another. What concerned him was not classification into species and genera, but rather the metamorphosis of natural phenomena as well as of individual beings in nature.

The way that all of post-Goethean natural science has gone into so-called natural causes was also not at all to Goethe's way of thinking. Concerning this point especially it is of great importance to become acquainted with the principal difference between the method of current natural science and the way Goethe approached nature. Current natural science conducts experiments. It investigates phenomena, attempts to elaborate them conceptually, and seeks to form notions of the so-called causes behind the phenomena—for example, the objective wave movement in the ether as the cause behind the subjective light and color phenomenon.

Goethe does not employ any of this style of scientific thinking. In his research he does not go from the so-called known into the so-called unknown at all. Instead he always wants to stay with the known, without at first worrying about whether the known is merely subjective—an effect on our senses, our nerves, our soul—or objective. Concepts such as subjective color phenomena or objective wave movement out there in space do not figure with Goethe at all. Instead what he sees revealed in space and taking place in time is something completely undivided whose subjectivity and objectivity he does not question. He does not employ the thinking and methods used in the natural sciences to induce the unknown from the known. Rather he employs all his thinking and all his methods to putting the phenomena themselves together, so that, by juxtaposing them, he finally arrives at phenomena he calls archetypal phenomena, which in turn, without consideration of their subjectivity or objectivity, express what he wants to make the basis of his study of nature and of the world. Therefore Goethe stays within the sequence of the phenomena; he merely simplifies them and then regards the simple phenomena that can be comprehended in this way as the archetypal phenomenon [*das Urphänomen*].

Thus Goethe regards the whole of what we can call the scientific method only as a tool for grouping the phenomena within the phenomenal sphere itself so that they reveal their own secrets. Nowhere does Goethe attempt to take refuge from a so-called known in any unknown. Therefore for him there is also nothing that we can call a natural law.

You have a natural law if I say that in their orbits around the Sun the planets make certain motions that describe such and such paths. For Goethe it was not important to arrive at such laws. What he expresses as the basis of his research are facts, for example, the fact of how light and matter placed in its path affect each other. He expresses the effect in words; it is not a law, but a fact. And he attempts to base his study of nature on such facts. He does not want to ascend from the known to the unknown. He also does not want to have laws. What he actually wants is a kind of rational description of nature. Only for him there is a difference between the initial description of the phenomenon, which is unmediated and complex, and the description gained by uncovering the simplest elements. Goethe uses these simple elements as the basis of his study of nature, in the same way that otherwise the unknown or the purely conceptually posited framework of laws is used.

There is something else that can cast light, so to speak, on the content of our natural sciences and on what is seeking to enter them through Goetheanism. Hardly anyone had such clear ideas as Goethe about the relationship of natural phenomena to the mathematical way of looking at things. Of course, this is always disputed. Simply because Goethe was not a crafty mathematician, people dispute that he had a clear view of the relationship of natural phenomena to the mathematical formulations that have become more and more popular, and are actually simply the safe thing in natural science today. The point is that the mathematical way of looking at natural phenomena (it would be false to call it the mathematical study of nature), the study of natural phenomena by means of mathematical formulations, has become standard for the way that we imagine nature.

We have to gain some clarity about these things. The usual path to understanding nature comprises three different kinds of approaches. People employ these three before actually arriving at nature itself. The first approach is ordinary arithmetic. In today's natural sciences we calculate to an extraordinary degree. We calculate and we count. Now we must be clear that arithmetic is something that people grasp purely through themselves. What we count when we count is a matter of complete indifference. By taking up arithmetic we are using something that at first blush has no relationship to the outer world at all; we could just as well be counting peas as electrons. The way of determining that our methods of counting and calculating are right is an entirely different matter from the results we see in the process to which we apply arithmetic.

There is a second approach that we practice before we arrive at nature itself. It is the way that we work with geometry. We determine what a cube or an octahedron is, and what their angles are, without extending our observations to nature. It is something we fabricate out of ourselves. The fact that we draw these things is only a function of our laziness. We could just as well simply imagine everything that we illustrate, and it is even useful if we just imagine some things and use illustrations less often as a crutch. It follows that what we express about geometric form is taken from a region that is initially distant from outer nature. We know what we have to express about a cube without deriving it from a cube of rock salt. However, the geometry must be found in the rock salt too. Thus we do something that is distant from nature and then apply it to nature.

A third approach, with which we still do not penetrate to nature, is what we practice in the science of motion, what is known as kinematics. Now kinematics is actually also something quite distant from the "real" natural phenomenon. You see, rather than looking at a moving object, I imagine the movement. I imagine that an object moves from, say, point *a* to point *b* [Figure 1a]. I even say that point *a* moves toward point *b*. I imagine it. I can also imagine this movement from *a* to *b* to be composed of two movements. Imagine for a moment that point *a* came to point *b*, but that it did not immediately move directly to point *b*. Instead it moved first to *c*. If it subsequently moves from *c* to *b*, it also arrives at *b*. Thus I can also imagine the movement from *a* to *b* such that it does not take place on the line *a-b*, but on the line or on the two lines *a-c-b*. That means I can imagine that the movement *a-b* is composed of *a-c* and *c-b*, in other words of two other movements. You do not have to observe a natural event at all. You can simply imagine that movement *a-b* is composed of the two other movements. That is, instead of one movement, two movements can be carried out with the same effect. Now, if I imagine this, it is a pure construct because, instead of drawing it, I could have given you instructions for visualizing the situation, and that would have to be a valid concept for you.

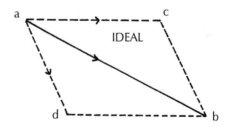

Figure 1a

However, if there really is such a thing in nature as point *a*, for example a single grain of shot, and it moves first from *a* to *b*, and another time from *a* to *c* and then from *c* to *b*, then what I have imagined really takes place. In other words, in kinematics

I imagine the movements, but for this concept to be applicable to natural phenomena it must hold for the natural phenomena themselves.

Thus we can say that in arithmetic, geometry, and kinematics we have three preliminary stages of the study of nature. The concepts we gain from them are pure constructs, but they are authoritative for what happens in nature.

Now I would like you to take a little walk down memory lane into your more or less distant study of physics and recall that you were once confronted with something called the parallelogram of forces [Figure 1b]: if a force acts on point *a*, this force can pull point *a* to point *b*. Now, by point *a* I mean something material—let's say a tiny grain. I pull it from *a* to *b* by means of a force. Please note the difference between what I am saying now and what I said before. Before I spoke of the movement. Now I am saying that a force pulls *a* toward *b*. If you express in line segments the measurement of the force, say five grams, that pulls from *a* to *b* (see illustration)—one gram, two grams, three grams, four grams, five grams—then you can say, I am pulling *a* to *b* with a force of five grams.

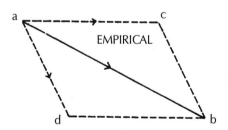

EMPIRICAL

Figure 1b

I could also arrange the whole process differently. I could first pull *a* to *c* with a given force, but, if I pull it from *a* to *c*, then I can still carry out a second pull. I can pull in the direction indicated here by the line connecting *c* to *b*, and then I have to pull it with a force that corresponds to this length. Thus, if I pull *a* to *b* with a force of five grams, I would be able to calculate based on this figure how large the pull *a-c* must be and how large the pull *c-b* must be. If I pull *a* toward *c* and *a* toward *d* at the same time, then I am still pulling *a* so that it will finally come to *b*, and I can calculate how strongly I have to pull *a* toward *c* and how strongly toward *d*. However, I cannot calculate this in the same way that I calculated the movement in the above example. What I determined above for the movement can be calculated as a concept. As soon as an actual pull, that is, an actual force, is applied, I have to measure this force somehow. Then I have to go to nature itself. I have to make the leap from the concept into the world of facts.

The clearer you become about the difference between the movement parallelogram—it is a parallelogram too if you add this point [*d* in Figure 1a]—and the parallelogram of forces, the more clearly and precisely you will express the difference between what can be determined conceptually and what lies beyond the reach of concepts. Conceptually you can arrive at movements, but not at forces. Forces have to be measured in the physical world. And only if you establish it externally by experimentation can you confirm that if two pulls are carried out, from *a* toward *c* and from *a* toward *d*, then *a* will be pulled to *b* according to the laws of the parallelogram of forces. There is no conceptual proof whatsoever as in the above example.

Therefore we can say that the movement parallelogram is derived by pure reason, while the parallelogram of forces has to be derived empirically through external experience. By distinguishing the movement parallelogram from the parallelogram of forces, you have the precise difference between kinematics and mechanics. Mechanics, which deals with forces, not merely with movements, is a natural science, whereas arithmetic, geometry, and kinematics are not. Only mechanics deals with the effects of forces in space and time. But one has to go beyond the world of concepts to arrive at this first natural science, mechanics.

Even on this point our contemporaries do not think clearly enough. I want to give you an example to illustrate what a mighty leap it is from kinematics to mechanics. The phenomena of kinematics can transpire completely within a conceptual space, whereas mechanical phenomena can at first be tested only in the physical world. People do not realize this clearly enough, so they are forever confounding things that we can understand mathematically with things in which entities of the physical world already come into play. For what is required whenever we speak of the parallelogram of forces? As long as we are speaking of the movement parallelogram, there need be nothing more than an imaginary body, but with the parallelogram of forces there has to be a mass, a mass that has weight, for example. That is something we have to realize: at *a* there must be a mass. Now you probably feel the urge to ask, "What is a mass actually?"

To a certain extent you will have to say, "Here I already falter."[4] For, as it turns out, whenever we depart from things that can be determined in the conceptual world so that they are valid for nature when we go into them, we are standing on fairly shaky ground. You know, of course, that in order to get by we equip ourselves, so to speak, with arithmetic, geometry, and kinematics, and the little bit that is brought in from mechanics. Then, by means of the mechanics of the molecules and atoms into which we believe so-called matter to be divided, we attempt to understand the natural phenomena that we initially experience subjectively. We touch a warm object. The natural scientist tells us that what we call heat is the effect on our heat nerves. What is objectively present is the movement of molecules and of atoms, which you can study according to the laws of mechanics. Thus we study the laws of mechanics of atoms and molecules, and we have long thought that by studying the mechanics of atoms, etc., it would be possible to explain all natural phenomena in general. Nowadays this idea is already beginning to waver. Even so, even if you penetrate conceptually to the atom, you have to inquire, by all sorts of experiments, how the force arises and how the mass acts. If you get as far as the atom, then you have to ask further how an atom can be recognized. To a certain extent you can recognize the mass only in its effects.

We have grown used to recognizing the smallest thing that we describe as a carrier of mechanical force by its effect.[5] Thus, we have answered the question by saying that if the smallest such piece of matter sets another piece in motion, say a small piece of matter weighing one gram, then a *force* must be exerted by that piece of matter which sets the other piece in motion. If this mass sets the other mass weighing one gram in motion, such that the other mass is accelerated one centimeter per second in a second, then the first mass has exerted a force that we have become accustomed to look upon as a sort of "universal unit." And if we can say that some force is so many times greater than the force that must be exerted to accelerate a gram one centimeter per second in one second, then we know how this exertion of force compares to a certain universal unit. If we were to express this universal unit in terms of weight, it would be 0.001019 gram [i.e., one dyne—Trans.]. Thus we would be able to say that such an atomistic body, whose exertion of force we do not investigate any further in nature, is capable of giving any body weighing one gram a shove that will accelerate it one centimeter per second in a second.

But how can we express what this force is made of? We can do it by going to the scales. This force is equal to the pressure that we read as 0.001019 gram on the scales. Thus I have to express myself in very real, external terms if I want to get to what we call mass in the world. I can express what I conceive of as mass by introducing weight into the situation—something I have gotten to know externally. I express the mass only in terms of weight. Even if I go into the atomization of mass, I express myself in terms of weight.

That is exactly the point I would like to describe: where we depart from what can be determined a priori and arrive at nature itself. I want you to understand to what degree the results ascertained apart from nature by means of arithmetic, geometry, and kinematics are usable. You should be clear to what extent they can be definitive for something that actually meets us on a completely different plane; it first meets us in the science of mechanics and can only then actually be the content of what we call a natural phenomenon.

Goethe recognized clearly that it is possible to speak of natural phenomena only when we pass from kinematics to mechanics. Because he knew this, it was very clear to him what the sole relevance of mathematics, which has been so idolized for the natural sciences, could be for this natural science.

I would like to clarify this with an example. Just as we can say that the simplest element in the exertion of natural forces would be any given atomistic body capable of accelerating one gram one centimeter per second in a second, we could also conclude that in all instances where force is exerted, the force emanates from a given point and acts toward a given point. Thus we could get into the habit—a habit that is quite the usual thing in the natural sciences—of searching more or less everywhere for points from which forces emanate. In numerous cases we will see that we have phenomenal fields and that we go back from these fields to the points from which the forces that dominate the phenomena emanate. Thus we speak of such forces whose point of origin is sought as central forces because they always emanate from centers. We could also say that we are justified in speaking of central forces whenever we come to a point where quite specific forces emanate that dominate a phenomenal field. But it is not always necessary for this play of forces to take place. It can also be the case that there is merely the potential for this play of forces to take place and that these forces will become active only if certain conditions arise in the surrounding area.

In the course of these days we will see how to a certain extent forces are concentrated in points without coming into play. Only if we fulfill certain conditions do they call forth phenomena in their surroundings. However, we have to understand that in a given point or a given space forces are concentrated that can act upon their surroundings. That is actually what we always find when we speak of the world in physical terms. All physical research consists of pursuing the central forces to their centers, of attempting to penetrate to the points from which effects can emanate. Thus we have to assume that there are centers for such natural effects that are charged, so to speak, with possible effects in certain directions. Indeed we can measure these possible effects by all sorts of procedures, and we can also express in measurements how strongly such a point can act. In general, when forces that can act when we fulfill certain conditions are concentrated in a given point, we call the measurement of the forces concentrated there the potential, the potential force. Thus we can also say that when we study natural effects, we are intent on pursuing the potentials of central forces. We go toward certain middle points in order to study them as the point of origin of potential forces.

This is basically the path taken by the particular direction of natural science that would like to transform everything into mechanics. It searches for the central forces, or better, the potentials of the central forces. But taking the important step into nature itself is a question of clearly realizing that you cannot understand a phenomenon in which life plays a role if you proceed only according to this method, if you only search for the potentials of central forces.   If you are studying the play of forces in an animal or plant embryo, you will never succeed. But in fact the ideal of modern natural sciences is to study organic phenomena through potentials, through central forces of some description. It will be the dawn of a new worldview in this discipline when we arrive at the realization that the pursuit of such central forces will not work to study phenomena in which life plays a role. And why not? Well, let's imagine for the sake of simplicity that we wanted to study natural processes by physical experimentation. We go to the centers and study the possible effects that can emanate from such centers. We find the effect. Thus when I calculate the potentials of the three points a, b, c, I find that a can affect α, β, γ; likewise, c can affect α^1, β^1, γ^1, etc. I would then get an idea of how the effects of a given sphere play out under the influence of the potentials of certain central forces. Using this method, however, I will never be able to explain anything in which life plays a role. Why? Because the forces that are involved in life do not have potentials and are not central forces.

Thus if you were to try in this case to find in d the physical effects under the influence of a, b, c, you would be able to go back to the central forces. If you wanted to study the effects of life, however, you could never say this, because there are no centers a, b, c for life effects. Instead you can understand the situation correctly only if you say, "In d I have life." Now I look for the forces that have an effect on life. I cannot find them in a, b, c, and not even if I go further, but only if I go more or less to the end of the universe, in fact, to its entire surroundings. In other words, starting from d, I would have to go to the end of the world and conceive that forces are acting inward from every point in the sphere, coinciding in such a way that they all come together in point d. Thus it is the complete opposite of central forces, which have a potential. How could I calculate a potential for something that acts from all sides from the infinity of space! It would have to be calculated by dividing the forces. I would have to divide a total force into smaller and smaller parts as I came closer to the edge of the world. The force would fragment. Every calculation would fragment too, because in this case universal forces, not central forces, are at work. That is where calculations cease. And that is once again the leap from lifeless nature into living nature.

We can find our way to a real study of nature only when we understand first the leap from kinematics to mechanics, and when in turn we understand the leap from outer nature to something that can no longer be arrived at through calculations because every calculation fragments and every potential disintegrates. By this second leap we pass from outer, inorganic nature to living nature. However, in order to grasp what life is, we must be clear how all calculations come to an end.

Now I have neatly separated out for you everything that can be traced from potential and central forces from that which leads to universal forces. However, out there in nature it is not separated in this way. You could pose the question, where is there a situation where only central forces act according to potentials, and where is there the other situation, where universal forces are at work that are not calculable according to potentials? There is an answer to this question, but it immediately indicates what important considerations have to be taken into account. We can say that in everything that people produce in the way of machines, which are put together from natural elements, we find purely abstract central forces according to their potential. Whatever is found in nature, however, even inorganic things, cannot be studied solely according to central forces. That does not exist. That does not add up. Rather, in every case, where we have to do with things that are not artificially produced by people, what we are dealing with is a confluence that takes place between the effects of central forces and the effects of universal forces. In the entire realm of so-called nature we find nothing that is lifeless in the true meaning of the word, with the exception of what people produce artificially—their machines, their mechanical products.

In a deeply instinctual way this was something that was both clear and unclear for Goethe, for it was an instinct on which he based his entire view of nature. And the contrast between Goethe and the natural scientist as represented by Newton actually derives from this fact—in modern times the natural scientist has studied only this one thing: the observation of outer nature solely for the purpose of tracing it back to the central forces and for driving out of nature everything that could not be determined by central forces and potentials. Goethe did not accept the validity of such an approach, for to him what was called nature was only a lifeless abstraction under the influence of this approach. For him there was something real only when, in addition to central forces, forces from the periphery, universal forces, come into play. Basically, his entire theory of color is also built upon this contrast. But we will come to speak about that in detail in the next few days.

I especially wanted to give you this introduction today so that you could understand the relationship of the human being to the study of nature. In our times we have to devote ourselves once again to a study like the one we have carried out today, because now the time has come when we have a subconscious glimmering of the impossibility of the modern approach to nature and some sense that things have to change. People still laugh a good deal when it is said that the old view of things does not work, but a time will come in the not-so-distant future when they will stop laughing, a time when we will be able to speak in Goethe's sense even about physics. Perhaps we will speak about color in Goethe's sense when another fortress that is regarded as even stronger is stormed, a fortress that even now has begun to crumble. That is the fortress of the theory of gravity. In this area especially, new theories emerge almost every year that shake the Newtonian conception of gravity, which relies purely on the notion that only the mere mechanism of central forces should figure.

I believe that especially today the teachers of youth, as well as those who want to have a hand in the development of culture, must create a clear picture for themselves of how the human being stands in relation to nature.

Second Lecture

YESTERDAY I SPOKE to you about how one side of natural science is the merely kinematic, which we achieve through the life of the imagination simply by forming concepts about all physical processes in terms of number, space, and movement. We are able to fabricate the kinematic, so to speak, whole cloth out of the life of the imagination. It is quite significant that the mathematical formulas we obtain concerning number, space, and movement do actually fit the natural processes themselves. On the other hand, it is equally significant that the moment we advance past number, space, and movement only as far as mass, we have to refer to outer experience.

Yesterday we explained this for ourselves and also gained from this the insight that modern physics has to make this leap from the inner reconstruction of natural events by kinematics into external sense experience without actually being able to understand the leap. You see, without taking steps to understand this leap, it will be impossible ever to gain a conception of what should be called the "ether" in physics. For example, as I pointed out to you yesterday, according to present-day physics, although it has started to become uncertain about these notions, light and color effects act upon us as sentient beings, as beings with nerves or even with souls, but these effects are subjective. What happens out there in space and time is objective movement in the ether. However, if you look into the literature of contemporary physics or elsewhere in the world of physics for the ideas that have been developed about this "ether," which supposedly creates the phenomena of light, you will find that these ideas are contradictory and confused and that you cannot get a proper idea about the "ether" with the tools modern physics has at its disposal.

We want to try to take the path that will bridge the chasm between kinematics and mechanics—for it is the latter, of course, that deals with forces and masses. I want to present a formula to you today just as a theorem; what it expresses will not occupy us until later, so those of you who may not recall it from your school years will be able to review what is necessary to understand it. I will put the elements together so that you can see this formula for a moment in your mind's eye.

Let's assume now, in accordance with the principles of kinematics, that a point (we always have to speak of a "point") moves in this direction. We are looking now only at the movement, not at its cause. Such a point will move either faster or slower, so we can say that it moves with greater or lesser velocity. Let's call the velocity v. Thus, this is a greater or a lesser velocity. As long as we do not pay attention to anything but the fact that such a point moves with a certain velocity, we remain within the bounds of kinematics. However, with such a notion we would not arrive at nature, not even purely mechanical nature. If we want to get to nature, we have to consider both what causes the point to move and the fact that a purely imaginary point cannot move—that is, if it is to move, the point must be something in external space. In short, we have to assume that a force acts on this point. I will call the velocity v and the force that acts on this point F.

Let's assume that this force does not push, so to speak, just once on this point in order to move it, causing it to fly off at a given velocity as long as it meets no obstacles. Instead let's begin with the assumption that this force acts continuously. In other words, the force acts on the point along its entire path. And let's call the distance along which this force acts on the point *d.* We also have to take into account the fact that the point must be something in space, and this *something* can be larger or smaller. Depending on whether this *something* is larger or smaller we can say that the point has a greater or smaller mass. For the moment we will express the mass in terms of weight. We can weigh what the force moves and express it in terms of weight. Let's then call the mass *m.*

Of course, if force *F* acts on mass *m,* a certain effect must take place. This does not manifest itself in the mass's having a constant velocity, but rather in its moving faster and faster. The velocity becomes greater and greater. In other words, we have to take into account that we are dealing with an increasing velocity. A smaller force acting on the same mass will be able to effect a smaller increase in velocity, while a larger force acting on the same mass will be able to effect a larger increase in velocity. Let's call this measure of the increase in velocity the acceleration and indicate it by the symbol *a.* And here I want to remind you of a formula that you probably already know, but should recall, for what interests us above all is the following: If you multiply the force that acts on the mass by the distance, you get a product equal to—that is, it can be expressed by—the mass multiplied by the square of the velocity divided by two. That is,

$$Fd = \frac{mv^2}{2}$$

Looking at the equation, you see that the mass is on the right side. You can gather from the equation that the bigger the mass is, the more force is required. However, what interests us now is that we have mass on the right side of the equation—the thing we can never arrive at through kinematics. Should we simply admit that everything lying beyond the bounds of kinematics has to remain forever inaccessible, so that we can only get to know it from staring at it, so to speak, from observation—or is there a bridge between kinematics and mechanics that modern physics cannot find? Modern physics is unable to find the transition point—and the consequences are appalling—because it has no real science of the human being, no real science of physiology. For in actuality we do not know the human being.

If I write v^2, I have something that has to do purely with number and movement. To that extent it is a kinematic formula. If I write m, I have to wonder if there is something in me that corresponds to m in a way similar to the way my conception of number and space corresponds, for example, to what I designate with v. What corresponds to m? What am I doing here actually? Physicists are normally not aware of what they are doing by writing m. This leads us back to the question, is there any way I can comprehend what is contained in m that is similar to the way I use kinematics to comprehend v? We can do this if we realize the following. If you press on something with your finger, you become familiar to an extent with the simplest form of pressure. Indeed mass reveals itself initially in no other way than in its being able to exert pressure. (As I have already told you, you can visualize mass by weighing it.) You can get to know such pressure by pressing on something with your finger. However, now we must wonder if something happens in us when we press on something—in other words, when we experience pressure—that is similar to comprehending a moving body. Yes, something of this kind does occur. You can understand what happens by making the pressure stronger and stronger. Just try—rather, it is better not to try—exerting pressure on a spot on your body and increasing the pressure, making it stronger and stronger. What will happen? If you make it strong enough, you will faint. In other words, you will lose your consciousness. You can infer from this that this phenomenon of loss of consciousness also takes place on a small scale, so to speak, if you only exert tolerable pressure. You just lose so little of the force of consciousness that you are still able to stand it. However, what I have characterized as a loss of consciousness under pressure so great that you cannot tolerate it is partially present on a small scale whenever we come somehow into contact with the effect of pressure—with the effect that emanates from a mass.

Now you only have to pursue this thought further, and you will not be far from understanding what we designate with *m*. While everything that is kinematic is unified with our consciousness in a neutral way, so to speak, we are not in this situation with that which is designated with *m*. Rather, with *m* our consciousness is instantly deadened. We can tolerate small doses of this deadening; large ones are beyond us. Fundamentally, however, in both instances it is the same thing. When we write *m*, we write something in nature that cancels our consciousness out when united with it—that is, it puts us partially to sleep. Thus we enter into a relationship with nature, but one that partially puts our consciousness to sleep. You see why that cannot be pursued kinematically. The kinematic is completely neutral to our consciousness. If we go beyond it, we enter areas that are opposed to our consciousness and cancel it out. Therefore, when we write the formula

$$Fd = \frac{mv^2}{2}$$

we have to say that human experience includes *m* as well as *v*, but that our ordinary consciousness does not suffice to comprehend *m*. This *m* immediately drains away the power of our consciousness. Here you have a real relationship to the human being—a completely real relationship to the human being. You see that states of consciousness have to be used in order to understand what is in nature. Without their help we will not even succeed in making just the step from kinematics to mechanics.

Nevertheless, even if we cannot live with our consciousness in anything that can be designated by m, for example, we do live in it with our whole selves as human beings. In particular, we live in it with our will, and we live very strongly in it with our will. Let me give you an example to illustrate how we live in m, in nature, with our will.

Once again I have to start out from something you will remember from your school years. I am going to recall something for you that you were well acquainted with during your school years. You know that, if we have a scale here [Figure 2a], and put a weight on it here, then take an equally heavy object, which I am just going to hang here, in order to balance the scale, then we determine the object's weight. The moment we place a vessel of water here, filled to here [see illustration], and immerse the weight in the water, the scale beam immediately rises. By being immersed in water the object becomes lighter, loses some of its weight. And, if we check to see how much lighter it has become—if we note how much we have to subtract to bring the scale into equilibrium once again—then we find that the object has lost a weight equal to the weight of the water it displaced. Thus weighing this volume of water gives us the loss of weight. You know that this is called the law of buoyancy, which states that a body in a liquid becomes lighter by a weight equal to the weight of the liquid it displaces. Therefore, as you can see, when a body is in a liquid, it strives upward, thus escaping to a certain extent from the downward pressure—the weight. In this way we are able to observe by objective, physical means something that has great significance in the constitution of the human being.

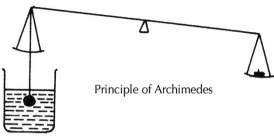

Principle of Archimedes

Figure 2a

On the average the human brain weighs 1250 grams. If the brain were actually to weigh 1250 grams when we carry it in ourselves, then it would press down so strongly on the blood vessels under it that it could no longer be properly supplied with blood. A heavy pressure would be exerted, which would instantly cloud our consciousness. In reality the brain doesn't press down on the base of the skull cavity with its full 1250 grams at all, but only with 20 grams. That is because the brain floats in the cerebrospinal fluid. Just as this body here floats in the water, the brain floats in the cerebrospinal fluid. And the weight of the cerebrospinal fluid that is displaced by the brain is equal to approximately 1230 grams. The brain becomes that much lighter and then weighs only 20 grams. That means that if we regard the brain as the tool of our intelligence and of our soul life, at least of a part of our soul life— as we indeed do with a certain amount of justification—then we should not be thinking only in terms of the weighable brain. For that is not the only thing there. Rather, by means of this buoyancy, the brain actually strives upward—strives upward against its own weight. That means that with our intelligence we do not live in forces that pull us downward, but rather in forces that pull us upward. With our intelligence we live in a state of buoyancy.

What I have explained to you applies only to the brain. The other parts of our organism, from the base of the skull down, are in this situation only to the smallest extent—with the exception of the spinal cord. But on the whole the other parts of the organism push downward. In the brain we live in buoyancy, striving upward; otherwise we live in the downward pull. Our will definitely lives in the downward pull. It has to unite with the downward pressure, but its consciousness is thereby taken away. Because of this it sleeps continually. That is precisely the essence of the phenomenon of will—that it is extinguished as a conscious phenomenon because it unites with the force of gravity, which is directed downward. And our intelligence becomes luminous because we are able to unite with buoyancy, because our brain works against the force of gravity.

You see, because of the different ways human life is united with the material basis, the submerging of the will in matter is effected on the one hand, and the enlightening of the will into intelligence on the other. Intelligence could never arise if our souls were bound solely to downward-pulling matter.

If we do not look at human beings in the abstract way of today, but look at them as they really are, we will actually experience that the spiritual comes together with the physical. But the spiritual must be conceived so strongly that it can also embrace physical knowledge—on the one hand, through a special union with material life, specifically with the buoyancy in material life, the enlightening into intelligence, and, on the other hand, through being put to sleep, when we have to let the will be drained away, so to speak, by downward pressure, making the will act in accordance with this downward pressure. This is the way that the will acts. Only a small part of it filters through to the area of 20-gram pressure and enters into intelligence. Thus, intelligence is imbued with will to an extent, but, in essence, with intelligence we are dealing with something that is opposed to ponderable matter. By thinking, we always strive beyond our heads.

Here you can see how physical knowledge must be merged with that which lives in the human being. If we remain within the bounds of kinematics, we are dealing with the abstractions that are so popular today, and we cannot build a bridge between them and the external reality of nature. We need knowledge with a spiritual content strong enough to really immerse itself in the natural phenomena, so that it can grasp, for example, how physical weight and buoyancy work within the human being.

Now I have shown you how human beings come to terms internally with the downward pressure and buoyancy—in other words, how they live into the interrelation between the kinematic and the material. But you see, for this we need a new scientific deepening. We cannot do it with the old scientific way of thinking, which invents wave movements or emissions that are nothing but pure abstractions. It seeks the path across to matter by means of virtual speculation, but naturally cannot find it in this way. A truly spiritual science seeks the path across into matter by trying to really immerse itself in matter, by pursuing will and intelligence in the soul life right down into the phenomena of pressure and buoyancy. Then you have real monism, which can arise only out of spiritual science—not the lip-service monism that is driven so strongly by ignorance these days. But it is especially necessary for physics, if you will excuse the expression, to get some smarts in its head, by making the connection between the phenomena out there and the physiological phenomenon of the floating brain. Once we have that connection, we know that is the way it must be, for Archimedes' principle cannot be invalid for the brain floating in cerebrospinal fluid.

Now, what happens, though, as a consequence of the fact that, with the exception of the 20 grams where the unconscious will comes into play, we actually live by means of our brains in the sphere of intelligence? To the extent that we use our brains as a tool, our intelligence is relieved of the burden of downward-pulling matter. This is eliminated to such a high degree that a weight of 1230 grams is lost. Matter is cancelled out to a very high degree. Through its being cancelled out to such a high degree, we are in the position to make our etheric body[1] work for our brain to a particular extent. The etheric body can do what it wants because it is not diverted by the heaviness of matter. In the rest of our organism the etheric is overwhelmed by the heaviness of matter. Thus the human being is organized in such a way that the etheric is free for everything that serves our intelligence. For everything else the etheric is bound to physical matter. Thus for our brain the etheric organism drowns out the physical organism, and for the rest of the body the arrangement and forces of our physical organism drown out those of the etheric organism.

I have already drawn your attention to the relationship you enter into with the outer world when you subject yourself to pressure. There is a narcotic effect. But there are also other relationships, one of which I would like to anticipate today—the relationship to the outer world that sets in when we open our eyes in a brightly lit room. Obviously in this situation an entirely different relationship to the outer world takes place than when we collide with matter and are introduced to pressure. Indeed when we expose ourselves to light, not only do we lose nothing of our consciousness, but, to the extent that light acts solely as light, any of us who so desire can feel how our consciousness takes an interest in the outer world when we expose ourselves to light, so that it virtually becomes more awake. The forces of consciousness unite in a certain way with what we encounter in the form of light; we will discuss that more precisely later. However, in light and under light we also encounter color. Light is actually something we cannot say we see at all. With the aid of light we can see color, but we cannot actually say we see light. Why we see so-called white light we will discuss later.

Everything that we encounter as color appears to us as a polar phenomenon, just as, say, magnetism appears to us as a polar phenomenon: positive magnetism, negative magnetism. In the same way we also encounter color as a polar phenomenon. One pole is everything that we designate as yellow, for instance, and related to yellow: orange and reds. The other pole is blue and everything that we designate as related to blue: indigo, violet, and even lesser shades of green. Why do I say that color appears to us as a polar phenomenon? If I may say so, the polarity of color must be thoroughly studied as one of the most significant phenomena in all of nature. If you want to get right down to what Goethe called the archetypal phenomenon, in the sense that I explained to you yesterday, you can arrive at the archetypal phenomenon of color first by searching for color in and under light.

Today, as a first experiment, we want to try as well as we can to seek out color in and under light. First, I will explain the experiment to you. We can do it as follows. We will let light pass through a small opening, which we will assume to be circular, cut into an otherwise opaque wall [Figure 2b]. We let the light flood in through this opening. If we set up a screen opposite this wall, the light pouring in causes a lighted circular area to appear. The best way to conduct the experiment is to cut a hole in the window shutter to let the light flood through and then set up a screen to capture the resulting image. We cannot do that here, but we can do it with the help of this projector by removing the cap. Then, as you can see, we get a shining circular area, which is initially nothing but the image that results because a beam of light produced here is captured on the opposite wall. Now we push a so-called prism into the path of this beam of light, preventing the light from simply reaching the facing wall and creating a circle there. Instead it is forced to deviate from its path. We cause that to happen by using a hollow prism, which is put together such that we have glass panes arranged in a wedge [Figure 2c]. This hollow prism is filled with water. We let the light beam created here pass through this water prism. Now if you look at the wall, you see that the disk of light is not at the same place down here where it was before. Instead, you see that it is raised—it appears at a different place. Besides that, you notice something else remarkable. Above, you see the edge in a bluish-green light, with a bluish-green edge, a bluish edge. Below, you see a reddish-yellow edge.

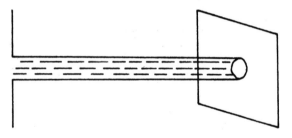

Figure 2b

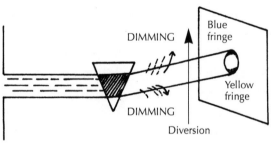

Figure 2c

There we have what we call a phenomenon. Let's hold on to this phenomenon for a moment. If we note down the facts, we have to note them thus: Somehow the light deviates from its path by going through the prism. It forms a circle up there. If we measured it, we would find that it isn't an exact circle. Rather it is lengthened at the top and bottom, and at the top it has a bluish edge and at the bottom a yellowish edge. Thus, you see, if we allow such a beam of light to pass through prismatically shaped water (we can ignore the changes caused by the glass plates), then color phenomena appear on the edges.

Now I am going to do the experiment once more with a much narrower beam of light. You now see a much smaller disk down here. If we divert this little disc by means of the prism, you see the spot of light up here, in other words once again displaced upward. However, now you see that the circle of light is almost completely taken up by colors. If I want to draw what you have here now, you see that what is displaced up here appears violet, blue, green, yellow, and red. Indeed, if we could investigate it completely, it would be arranged in perfect rainbow colors. Please, we are considering the facts, and I ask all of you who in school studied all the beautiful drawings of rays of light, of angles of incidence and so forth, to please forget them and to stick with the pure phenomenon, the pure fact. We see colors arise in and under light, and we can wonder why it is that such colors arise. If I put in the big circle once again, we have the light beam passing through space and falling there on the screen, where it forms an image. If we once again place the prism in the path of this light beam, then we get the deviation of the image and, in addition, the color phenomena on the edges.

Now, however, I would like to ask you to observe the following. We will stay purely within the facts. I am asking you to observe that, if you just look around a bit, you see the shining water cylinder right there as the light travels through the glass prism. The light beam passes—this is purely factual—through the water prism, and the light and the water fit into each other. Please pay close attention to this now. When the light beam passes through the water prism, the light and the water fit into each other. The light and water do not fit into each other without having an effect on the environment. Instead, we have to say that the light beam passes and—staying within the facts, as I said—somehow has the force to penetrate the prism and reach the other side. However, the prism diverts the light beam. It would go straight, but it is raised and diverted, so that we have to realize that something is present that diverts our light beam. If I want to indicate with an arrow what diverts our light beam, I would have to do it with this arrow. Now we can say— staying purely with the facts and not speculating—that the light is diverted upward by such a prism, and we can indicate the direction of the diversion.

Now I ask you to think along the following lines, which once again just correspond to the facts. If you allow light to pass through semiopaque milk glass or through any semiopaque liquid—in other words, through semiopaque matter— the light will naturally be weakened. If you look at light through clear water, you see its brightness; in semiopaque water it appears weakened. You can observe in countless cases how light is weakened by a semiopaque medium. We have to declare this as a fact. In any case, however, every material medium— even this one here [Figure 2c] that is acting as a prism—is semiopaque. It darkens the light. In other words, with the light within the prism we are dealing with a darkened light. On the left here we are dealing with shining light. On the right here we are dealing with light that has made its way through the medium. However, here inside the prism we are dealing with a combination of light and matter—with the origin of this darkening. You can gather that a darkening is taking effect simply by the fact that when you look at light through a semiopaque material you also see something else. Thus, a darkening takes place—this is perceptible.

What arises because of this darkening? We are not dealing simply with a light cone that turns aside and continues, but also with what comes into play as a darkening of the light, caused by matter. We can conceive of this as follows: it is not only the light that is radiating into this space beyond the prism, but it is also the turbidity within the prism that is radiating into there. How? It spreads out, naturally, after the light has passed through the prism. The turbidity radiates into the brightness. And you only need to think about the matter in the right way to say to yourself that the turbidity shines upward, that when the brightness is diverted, the turbidity is also diverted upward. That is to say, the turbidity is diverted upward in the same direction in which the brightness is diverted. Turbidity is relayed, so to speak, into the brightness that has been diverted upward. The brightness cannot just spread itself out up there; turbidity is relayed into it. We are dealing with two collaborators, with the diverted brightness and with the relaying of turbidity into this brightness, but the diversion of turbidity happens in the same direction as that of the brightness. You can see the result: because turbidity shines upward into the brightness, the dark and bluish colors arise.

Down below, what is it like there? Naturally turbidity is also shining downward. However, as you see, while one part of the radiating light is up here, with the turbidity going in the same direction as the surging light, we have here a spreading out of what arises as turbidity so that it shines in, and there is a space in which the light, for the most part, is diverted. But turbidity radiates into this body of light that has been diverted upward; here we have an area where turbidity passing through the upper areas of the prism goes downward. Because of this we have down here an area where turbidity is diverted in a way that is opposed to the brightness. We can say that here we have turbidity that wants to go into the brightness; but in the lower part the brightness is such that in its diversion it works counter to the diversion of the turbidity. The result is that, on the one hand, up here the diversion of turbidity takes place in the same way as that of the brightness; they both collaborate to an extent, and turbidity intervenes like a parasite, so to speak. On the other hand, down here turbidity radiates back into the brightness but is overwhelmed by it and suppressed to a degree. Thus the brightness dominates here and also dominates in the battle between brightness and turbidity. The results of this confrontation and of the brightness shining through the turbidity are the red or yellow colors down below. Thus we can say that up here turbidity courses into brightness, and blue tones arise; down below brightness drowns out the turbidity or darkness that is radiating in, and yellow tones arise.

You see that here, simply because the prism diverts the complete cone of light on the one side and the turbidity on the other side, in the two directions we are dealing with different ways in which darkness plays into brightness. We have an interplay of darkness and lightness, which do not mix to become gray, but instead remain independently effective. But toward the one pole they remain effective in such a manner that the darkness works toward the brightness to a degree, thus working to carry weight within the brightness, but as darkness. On the other side, the turbidity resists the brightness, remaining an independent presence, but is drowned out by the brightness. There the bright, yellowish colors arise. By staying purely with the facts and taking what is there purely from observation, you have the possibility of understanding why the yellows appear on the one side, and the blues on the other; and at the same time you can conclude that the material prism plays an important role in the emergence of the colors. Indeed, it is because of the prism that on one side the turbidity is diverted in the same way as the cone of light, while on the other side what is radiating forth and what is diverted cross, because the prism also lets its darkness radiate out to the other side, where a diversion has already taken place. Because of this, the diversion downward takes place, and darkness and brightness collaborate differently downward than upward. Thus colors emerge wherever darkness and brightness work together.

That is what I wanted to make especially clear to you today. If you want to consider now the best tack to take in order to understand this—then you have only to consider that your etheric body is engaged differently in the muscle than in the eye. In the muscle it connects itself with the functions of the muscle; but because the eye is very isolated, the etheric body is not engaged there in the physical apparatus—rather it is relatively independent. Because of this the astral body is able to achieve an intimate connection with the part of the etheric body in the eye.[2] Within the eye our astral body is independent in quite a different way than in the rest of our physical organization. Assume that this [drawing on the board] is a part of the physical organization in a muscle, and that this is the physical organization of the eye. If we describe it, we have to say that our astral body is engaged here as well as there; but there is a considerable difference. There it is engaged in such a way that it passes through the same space as the physical body, but not independently. Here, in the eye, it is also engaged, but it acts independently. Both fill out space in the same way, but in one instance the ingredients act independently, while in the other they do not act independently. Therefore, if we say that our astral body is in the physical body, it is only half of the story. We have to ask how it exists there, for it exists differently in the eye than in a muscle. Despite being in the eye just as it is in the muscle, the astral body is relatively independent in the eye. Thereby you can conclude that ingredients can penetrate each other and yet remain independent. Thus you can unite brightness and darkness to form gray; then they interpenetrate like astral body and muscle. Alternatively, they can penetrate each other such that they remain independent; then they inter-penetrate like our astral body and the physical organization of the eye. In the first instance gray emerges; in the second instance, color. If they penetrate each other like the astral body and muscle, gray emerges, and if they penetrate each other like the astral body and the eye, then color emerges, because they remain relatively independent, despite being in the same space.

Third Lecture

STUTTGART, DECEMBER 25, 1919

I HAVE BEEN TOLD that the culmination of our study yesterday, the phenomenon that appears by means of the prism, was difficult for many to understand, but please don't let that worry you. Understanding will come bit by bit. We will deal especially with light and color phenomena in more depth, so that this veritable pièce de résistance—for that is what they are even for the rest of physics—can provide us with a good foundation. You understand that first I have to tell you especially about some of those things that you cannot find in books and aren't the object of normal scientific studies—things that, to a certain extent, we can only deal with here. Then, in the last lectures, we can delve into how what we have studied here can also be used in lessons.

What I tried to explain to you yesterday is indeed essentially a special form of the interaction of brightness and turbidity. I wanted to demonstrate that color phenomena behaving as polar opposites to each other arise by means of the different kinds of interaction between brightness and turbidity, appearing especially during the passage of a beam of light through a prism. First I will ask you to swallow the bitter pill that the difficulty in understanding this matter comes from the fact that you (I am speaking of those who have difficulty in understanding) would actually like to have the light and color theory formulated kinematically. By virtue of our peculiar education, people have become accustomed to believe only those concepts that are more or less kinematic with reference to outer nature, that is, those concepts that deal only in terms of number, formal space, and movement. Now you are supposed to make the effort to think in terms of qualities, and really in a certain sense you can say, "Here I already falter." But you can ascribe that completely to the unnatural course scientific development has taken in more recent times, which you will even go through in a certain way with your students. (I mean the teachers of the Waldorf School[1] and other teachers.) For naturally it will not be immediately possible to inject healthy ideas into the school of today. Instead we will have to build bridges.

Now let's take a look at light and color phenomena from another point of view. I would like to begin today with a much-contested remark of Goethe's. In Goethe you can read how in the 1780s he became known for all sorts of assertions about the appearance of color in and under light,[2] in other words, about the phenomena that we started to talk about yesterday. He had been told that the general view of physicists was that if colorless light passed through a prism, it would be split, taken apart. The phenomena were interpreted more or less thus: If we catch a colorless beam of light, at first it reveals a colorless image to us. If we put a prism in the path of this beam of light, we get the series of colors red, orange, yellow, green, blue—light blue, dark blue—violet. This came to Goethe's attention; indeed he discovered that physicists explained the matter by saying that the colorless light actually already contains these seven colors—how is of course difficult to imagine, but this is what was said. When we pass the light through the prism, it does nothing but separate out, like a fan, what the light already contains. It breaks this down into the seven colors.

Goethe wanted to get to the bottom of this matter and borrowed all sorts of instruments, just as we have tried to collect them in the past few days, in order to determine how things were. He had Büttner, the Court Counselor in Jena, send him these instruments. Then he stored them away, intending to do his research at an opportune time. Counselor Büttner became impatient and demanded the instruments back before Goethe had done anything with them (with some things it just happens that we don't get to them right away). Goethe had to pack up the instruments, so he quickly took the prism, thinking that if light is broken up by a prism he would take a look at it on the wall. He expected that the light would appear neatly divided into seven colors—but color appeared only where there was some kind of edge, where there was a muddy spot on the wall, so that the muddiness, the darkness, collided with the brightness. There he could see colors if he looked through the prism, but where there was an even white he saw nothing. Then Goethe became suspicious and lost his faith in the whole theory. Now he had no desire to send back the instruments; he kept them and pursued the matter further. And it turned out that the matter is not at all the way it is ordinarily portrayed.

If we let light pass through the space of the room, we get a white circle on a screen. Now, if we put a prism in the path of this body of light that is passing through, the beam of light is diverted [Figures 2b, 2c]. Initially, however, the seven colors in a row don't appear at all. Rather, on the lower edge only a reddish color appears that merges into yellowish, and on the upper edge a bluish color that merges into greenish. In the middle it remains white.

What did Goethe say to himself then? He said, It really isn't a case of the light being split up. Instead, I am actually reproducing an image. This image is only the reproduction of the section here. The section has edges, and the colors don't appear because they are extracted from the light, because the light was split into them, but because I am projecting the image, and the image itself has edges. Thus here I am dealing with the fact that at the point where light and dark meet (for outside this circle of light there is darkness in the surrounding area and within it is light), there at the edges, the colors appear. The colors initially appear purely as marginal phenomena, and, by showing the colors as marginal phenomena, we basically have the archetypal phenomenon before us. We don't have the archetypal phenomenon before us at all if we make the circle smaller and get a continuous color image. While in the case of the big circle the marginal colors remain marginal colors, the continuous color image emerges only because in the case of the small circle the colors continue from the edge inward to the middle. They overlap in the middle and create what we call a continuous spectrum. Thus the archetypal phenomenon is the one in which colors appear on the edges, where light and dark merge.

You see, it is a matter of not using theories to interfere with the facts, but of staying purely with a study of the bare facts. We are not only concerned with what we see in terms of the appearance of colors. As you have seen, a displacement of the entire beam of light takes place—a sideways diversion of the entire beam of light. If you want to pursue this sideways diversion schematically, you can do so roughly as follows.

Assume that you fit two prisms together, so that the lower one, which together with the upper one forms a whole, is placed like the one I sketched for you yesterday [Figure 3a]. The upper prism is placed opposite the lower one. If I allowed a beam of light to pass through this double prism, I would naturally have to get something similar to what I got yesterday. I would get two diversions, one downward and the other upward. If I had such a double prism here, I would get a light image that was more extended lengthwise, but at the same time this more extended light image would turn out to be very indistinct and dark. This is understandable when, by capturing this image on a screen, I get a reproduction of the circle of light that is pushed into itself. But I could also move the screen in. Once again I would get a reproduction. In other words, there would be a distance—all of this remains within the facts— within which I would always find it possible to get a reproduction. You can conclude that we are manipulating the light by means of the double prism. Outside I will always find a red edge, in fact at both the top and the bottom, and violet in the middle. While otherwise I got only an image from red to violet, now I get red on the outer edges and violet in the middle, with the other colors in between. Thus with such a double prism I could make it possible for such a figure to emerge. However, I would also get this figure if I moved the screen. Therefore I have a certain distance within which it is possible for an image to appear that has color on the edges, but also has color in the middle and all sorts of transitional colors.

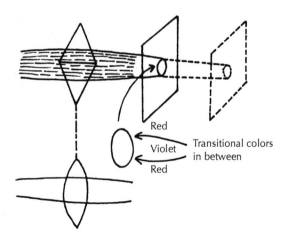

Red

Violet

Transitional colors
in between

Red

Figure 3a

Now if I walk up and down with the screen, we can prevent the possibility of such images being created in quite a wide space. However, you probably suspect that this possibility could only be created if I changed the prism constantly, because a prism whose angle is greater here would project the image at a different place than if I made the angle smaller and got a shorter distance here. I can make the whole situation different if I don't have a prism with straight surfaces, but instead work with curved surfaces from the very beginning. Thereby we can simplify considerably something that is otherwise extraordinarily difficult to study with a prism. And we get the following possibility: First we allow the beam of light to pass through the space, and then we place the lens, which is actually nothing but a double prism with curved surfaces, in the path. Now the image I get is considerably reduced in size. What has actually happened here? The entire beam of light is contracted, narrowed. Here we have a new interaction between matter, the matter in the lens, in the glass body, and the light that is passing through space. The lens acts on the light so that it contracts the beam of light.

We will sketch the whole thing out schematically. Here I have a beam of light, in a side view, and let the light pass through the lens. If I were to set up an ordinary glass or water plate against it, the beam of light would simply pass through it and a reproduction of the beam of light would result on the screen. That isn't the case if instead of a glass or water plate I have a lens. If I simply trace what has happened with a line, then I have to say that a reduction of the image has resulted. Thus the beam of light has contracted.

There is yet another possibility, which is that we don't copy the arrangement of a double prism like the one I have drawn here, but instead have a double prism whose cross section is formed such that the prisms meet at this edge [Figure 3b]. Of course, then I would get the same description I made before, but with a considerably enlarged circle. Once again, by walking up and down with the screen, I would have the possibility of getting a more or less distinct image within a certain distance. In this case I would have violet and bluish color above and also below, and in the middle I would have red. Before it was reversed. And in between are the intermediate colors.

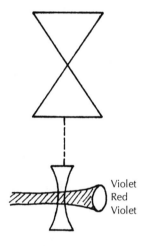

Violet
Red
Violet

Figure 3b

Once again I can replace this double prism with a lens having this cross section [Figure 3b]. This lens [Figure 3a] is thick in the cross section across the middle and thin at the edges. And this lens [Figure 3b] is thin in the middle and thick at the edges: in this case I get an enlarged image, which is significantly larger than the normal cross section that would emerge from the beam of light. I get an enlarged image, but also with the gradation of colors on the edges and toward the middle. Thus if I want to investigate these phenomena, I have to say that the beam of light has been expanded—it has essentially been driven apart. That is a simple fact.

What can we conclude from these phenomena? We see that there is a relationship between the substance we first encounter in the lenses as transparent matter and what makes its appearance because of the light. We also see a certain kind of interaction, for, given what we arrive at by using a lens that is thick at the edges and thin in the middle, what do we have to say when we have such a lens before us? That the whole beam of light has been driven apart—it has been expanded. And we also see how this expansion is possible—it takes place because the substance through which the light has passed is thin here and thicker here. There the light has to penetrate through more material than here in the middle, where it penetrates through less material. What happens then with the light? Well, we said that it is expanded, driven apart. It is driven apart in the direction of these two arrows. What can have driven it apart though? Well, simply the circumstance that it had to pass through less matter in the middle and more at the edges. Consider the situation: in the middle the light has less substance to pass through, and thus goes through more easily. Therefore when it has passed through, it still has more force. Here, where it goes through less material, it has more energy than where it goes through more material. The stronger force in the middle, which is caused by the light's going through less material, pushes the beam of light apart. This is something that you can infer directly from the facts.

Please be clear that this is a matter of the right use of the method, of logical thinking. If we draw lines to follow what appears in the light, we have to be clear that we are actually only adding something with the drawing that has nothing to do with the light. If I draw the lines here, then I am merely drawing the limits of the beam of light. This opening creates this beam of light. Thus what I am drawing has nothing to do with the light. Instead the beam is caused by the fact that the light passes through this slit. And if I say in this instance that the light moves in this direction, again that has nothing to do with the light, for, if I were to push the source of light up, then the light would move this way when it fell through the slit, and I would have to draw the direction of the arrows thus. All of this would have nothing to do with the light itself. We are accustomed to drawing lines into the light, so that we have gradually come to speak of rays of light. We aren't dealing anywhere with rays of lights. We are dealing with a cone of light that is caused by a slit through which we let the light pass. We are dealing with a broadening of the cone of light, and we have to say that somehow the broadening of the cone of light must be connected to the shorter path that the light travels here in the middle than here at the edge. Because of the shorter path here in the middle, the light retains more force. Because of the longer path at the edge, more of its force is taken away. The stronger light in the middle presses the weaker light at the edge, and the cone of light is broadened. That is what you can conclude.

Now you can see that, although we are actually dealing only with metaphors, in physics people talk about everything possible, about rays of light and such things. These rays of light have actually become the subsoil for materialistic thinking in this area. In order to make what I have just explained more concrete, we will look at something else. Assume that we have a tub here, a small container. In this small container we have a liquid, for example, water, and an object, let's say, a coin or the like, is lying on the bottom. If I have an eye here, I can do the following experiment: First, I can leave out the water and look at the object with the eye [Figure 3c]. I will see the object in this direction. What are the facts of the matter? I have an object lying on the bottom of a container. I look, and I see this object in a certain direction. Those are the simple facts of the case.

Now I begin to draw: a ray of light starts out from this object, is sent into the eye, and affects the eye—and then I fantasize all sorts of possible things about it.

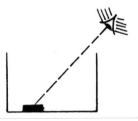

Figure 3c

Now I fill the container up to here with water or any other liquid [Figure 3d]. And something quite special happens. I trace the same direction from the eye to the object in which I saw the object before, and look in that direction. I could expect to see the same thing, but I don't. Instead, something highly peculiar occurs: I see the object a bit raised.[3] I see it in such a way that it is raised along with the entire bottom of the container. Of course, we can talk about how we can determine that, I mean measure it, later. Right now, I just want to talk about the principles. What could be the basis of this, if I am going to answer the question about the facts of the matter?

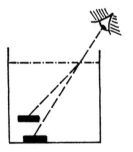

Figure 3d

Now I expect to find the object in the same direction in which I looked earlier. I direct my eye toward it, but I don't see it in that direction. I see it in another direction. Of course, before, when there was no water in the container and only air between my eye and the bottom, I was able to look down directly to the bottom. Now my line of sight collides with the water here, which doesn't let my eyesight through so easily as the air; it offers greater resistance, and I have to shy away from the greater resistance. From this point on I have to shy away from the greater resistance. This shying away is expressed in the fact that I don't see to the bottom, but that instead the whole thing appears to be raised. I see with more difficulty, so to speak, through the water than through the air. It is harder for me to overcome the resistance of the water than that of the air, so I have to shorten the force, thus pulling the object itself upward. I shorten the force in meeting a stronger resistance. If I were able to fill this with a gas that was thinner than air, the object would sink, because then I would meet less resistance. Thus I would push the object downward [Figure 3e].

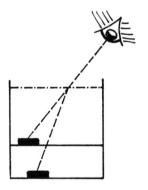

Figure 3e

Physicists don't state the facts of the matter in this way. Instead they say that a ray of light is thrown onto the surface of the water. The ray of light is bent there; because a transition takes place between a thicker medium and a thinner one, the angle of incidence of the ray of light is bent, and the light arrives at the eye here. And then they say something very curious: after the eye has received the information by means of the ray of light, it lengthens the path outward and projects the object onto this spot here. In other words, physicists discover all sorts of concepts, but they don't reckon with what is there, with the resistance that the power of vision itself meets in the thicker substance it has to penetrate. To a certain extent, they would like to leave out everything else and attribute everything to the light, just as in the case of the prism, where they say, "Oh, the prism doesn't do anything at all. The seven colors are already contained in the light. The prism only provides the occasion for them to line up so nicely next to each other like soldiers, the seven colors. But these seven naughty boys who are forced to step out separately are already in there together. The prism doesn't do any of it." As we saw, it was exactly what took place in the prism, in this darkened wedge, that caused the colors. The colors themselves have nothing to do with the light itself.

Here once again we have to be clear that we are actively carrying out work, aiming with our eye and meeting stronger resistance in the water, and because of this we are forced to shorten the line of sight through the stronger resistance. The physicists, however, say that rays of light are cast, bent, and so forth. And then here is the best part of all! Today's physicists say that first the light arrives at the eye on a bent path, and then the eye projects the image outward. What does that mean? In the end the physicists say, "The eye projects." They posit a kinematic conception, a conception bereft of all reality, a pure fantasy activity, in place of what immediately presents itself: the resistance of the thicker water to the eye's power of sight. It's at just such points that you notice most clearly how abstracted everything is in our physics, how everything is supposed to become kinematics, how they don't want to go into qualities. On the one hand, they divest the eye of any kind of activity; on the other hand, the eye projects outward the stimulus it receives. What is necessary, however, is that we begin at the outset with the activity of the eye, that we be clear that the eye is an active organism.

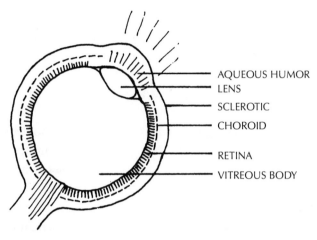

AQUEOUS HUMOR
LENS
SCLEROTIC
CHOROID
RETINA
VITREOUS BODY

Figure 3f

Now you see we have a model of the eye, and today we will also begin to deal with the nature of the human eye. The eye is of course a ball of sorts, just compressed a bit from front to back, a ball that sits here in the bone socket in such a way that first a series of skins surround the inner part of the eye. If I want to draw the cross section, I would have to do it like this [Figure 3f]—what I'm drawing now would be the right eye. If we were to take the eye out of the skull and dissect it, the outermost part, which we would find first, would be connective tissue, fat. But then we come to the first actual covering of the eye, the so-called sclera and cornea. This outermost covering is sinewy, bony, and cartilaginous. I've drawn it in here. It becomes transparent toward the front, so that the light can penetrate from here into the eye. A second layer lining the interior space is the so-called choroid, which contains the blood vessels. That would be approximately here. Third, we have the innermost layer, the so-called retina, whose continuation into the skull is the optic nerve. Thus the optic nerve would go inside here and form the retina. And here we've listed the three coverings of the eye. Now, however, behind the cornea, embedded here in the ciliary muscle, is a kind of lens. It is carried by a muscle here called the ciliary muscle. Toward the front here is the transparent cornea, and between it and the lens is what is called the aqueous humor. When light enters the eye, it first passes through the transparent cornea, then through the aqueous humor, then through this lens, which can be moved independently by muscles. Then the light passes beyond this lens into what we usually call the vitreous body, which occupies the entire space within the eye. Thus the light passes through the transparent cornea, the aqueous humor, the lens itself, the vitreous body, and from there to the retina, a branching of the optic nerve, which leads then into the brain. These things show us schematically the principal elements in the eye.

However, this eye reveals extraordinary peculiarities. First, if we study the aqueous humor, this liquid between the lens and the cornea through which light has to pass, we find that in terms of its contents it is almost a real liquid, almost an external liquid. At this point of the aqueous humor, between the lens and the outer cornea, the human being is physically like a piece of the outer world to a certain extent. It is almost the case that this liquid in the outermost periphery of the eye hardly distinguishes itself from a liquid I would pour onto my hand here. And the lens is also something very, very objective, very, very inorganic. However, if I go on to the vitreous body, which occupies the interior of the eye and borders on the retina, I cannot regard it in this way at all; I cannot say that this is also something almost like an external liquid or an external body. There is vitality in it already; there is life in it. Thus the farther we go back into the eye, the more we press forward into life. Here in the front we have a liquid that is objectively almost entirely external—even the lens is external—but in the case of the vitreous body we are already standing within a structure that has inherent vitality. This difference between what is outside and that which is inside shows itself in something else as well. Even this could be studied in the natural sciences today. If we look at the formation of the eye comparatively, starting with the lower order of animals, we find that what comprises the external aqueous body and the lens does not grow from inside out. Instead it is formed by the accretion of the surrounding cells. Thus I would have to think of the lens forming in such a way that the lens tissue and even the frontal aqueous humor emerge from the neighboring organs and not from the inside out, while the vitreous body grows from the inside toward them. Here we have something curious. The nature of external light is at work and brings about the transformation that produces the aqueous humor and the lens. The animal reacts to this from within and pushes something more alive, something more vital, out toward it—the vitreous body. It is particularly in the eye that the formations stimulated from the outside in and those stimulated from the inside out meet each other in quite a curious way. That is the first peculiarity of the eye.

There is yet another peculiarity, which lies in the fact that the spreading retina is actually the spreading optic nerve. The peculiarity consists in the fact—tomorrow I will try to show you an experiment to confirm this—that the eye is insensitive where the optic nerve enters. It is blind there. The optic nerve spreads out. At a place located somewhat to the right of the point of entry (in the case of the right eye), the retina is the most sensitive. Now you can say it is the nerve that senses the light. However, it doesn't sense the light precisely at its point of entry. You would think that if it were really the nerve that senses the light, it would have to sense it most strongly at the place where it enters. However, it does not. For the time being, I would like you to keep that in mind.

Now from the following example you can conclude that the setup of the eye is full of the extraordinary wisdom of nature. As long as your eyes are more or less healthy, you find that when you examine the objects around you during the day they appear more or less sharp to you. In any case their sharpness or clarity is sufficient for orientation. But sometimes when you wake up in the morning, you see the edges of objects very unclearly, as if they were surrounded by a little fog. If the object is a circle, you see something unclear around it. What is the cause of this? It is caused by the fact that we have three different things in our eye. We will take just two into consideration—the vitreous body and the lens. As we have seen, they have completely different origins. The lens is formed more from outside, the vitreous body more from within. The lens is less alive; the vitreous body is permeated with vitality. The moment we awake, the two of them are not yet adjusted to each other. The vitreous body tries to reproduce the objects the way it can, and the lens does it the way it can. And we have to wait until they have mutually focused themselves. You can see from this how mobile the organic is and how the effects of the organic are based on the fact that the activity in the lens and that in the vitreous body are differentiated at first and are then put together again out of the differentiated elements. They both have to adjust to each other.

On the basis of all these things, we will try little by little to arrive at how the world of color results from the interrelationship between the eye and the outer world. To this end, we want to take a look at the following experiment now so that tomorrow we can connect it to observations about this relationship of the eye to the outer world.

As you can see, here I have a disk painted with the colors that have come to our attention as the colors of the rainbow: violet, indigo, blue, green, yellow, orange, and red. If you look at the wheel here, you see these seven colors. I have done it as well as it is possible to do with these colors. Now first we'll turn the disk. You still see the seven colors, just in movement. Now I will make the disk rotate very rapidly. When the thing rotates fast enough, you don't see the colors anymore. Instead, you see, I believe, a monochromatic gray. Isn't that right? Or did you see something else? [Audience replies: "Purple." "Reddish."] Yes, that's because the red is somewhat too strong in relation to the other colors. I did try to compensate for that by giving it less space, but if the arrangement were completely right, you would actually see a monochromatic gray. Then we have to wonder why these seven colors appear to us as monochromatic gray. We'll try to answer this question tomorrow.

To day we'll just put forth what physics says. It says, just as it said in Goethe's time, that I have the colors red, orange, yellow, green, blue, indigo, and violet. Now I rotate the disk. Because of this, the impression of the light does not have an effect on the eye. Instead, if I have just seen red here, because of the rapid rotation of the disk, orange is already there, and if I have seen orange, yellow is already there, and so forth. And while I still have the other colors, red is already there again. Because of this I have all the colors at the same time. The impression of red is not yet finished when violet arrives. Thereby the colors are put together for the eye, and there has to be white once again. This was also the teaching in Goethe's time. Goethe was taught that if we make a color disk and rotate it quickly, the seven colors, which are so well behaved when they separate from each other out of a beam of light, will unite once again in the eye itself. However, Goethe never saw white. Rather he said that all you ever got was gray. Of course, more recent physics books also find that you only get gray. However, in order to make the story turn out white after all, they advise that a black contrast circle should be made in the middle. Then the gray will appear white in contrast. This, as you can see, is a nice way of doing it. Some people do it with Fortune; the physicists do it with nature. Thus do we correct nature. It actually happens that nature is corrected on a number of the most fundamental facts.

You see, I'm trying to proceed in such a way that a foundation is laid. Only when we have laid a good foundation will we be able to make progress in all areas.

Fourth Lecture

UNFORTUNATELY, we haven't gotten far enough along with putting together the materials for the experiments. Therefore tomorrow we will do some things that we wanted to do today, and I will have to adapt today's lecture so that I still present something that will be useful to us in the coming days, with only a slight change in my intentions, more or less.

First, I would simply like to describe what could be called the archetypal phenomenon of the theory of color. It will then be a matter of seeing this archetypal phenomenon gradually confirmed in the phenomena you can observe over the entire range of so-called optics or color theory. Of course, the phenomena become more complex, and the simple phenomenon doesn't immediately reveal itself so easily everywhere. However, if we make the effort, we will find it everywhere. This simple phenomenon, stated initially in the Goethean way, is as follows: If we see something brighter through darkness, then the brightness will tend toward the bright colors, in other words toward yellows or reds. If, for example, I see any shining, so-called whitish-appearing light through a sufficiently thick sheet of glass that has been darkened somehow, then what I see as whitish when I look at it directly now appears to be yellowish—reddish yellow [Figure 4a]. Brightness seen through darkness appears yellow or reddish yellow. That is one pole. Conversely, if you simply have a black surface here and you look at it directly, then you see just the black surface. Let's assume, however, that I have a trough of water here, and I shoot light through it so that it is illuminated. Then I have an illuminated liquid here, and I see the darkness through the brightness. I see it through something illuminated. Then blue or violet (purple) appears—in other words, the other pole of color [Figure 4b]. That is the archetypal phenomenon: brightness through darkness—yellow; darkness through brightness—blue.

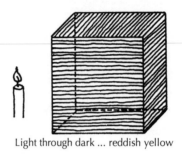

Light through dark ... reddish yellow

Figure 4a

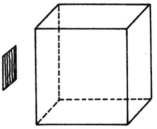

Dark through something illuminated ... blue-violet

Figure 4b

This simple phenomenon can be seen everywhere if we just get used to thinking concretely instead of abstractly, the way modern science thinks. With this in mind, recall the experiment we already conducted, where we let a beam of light pass through a prism and thereby got a true spectrum of colors from violet to red, which we captured. I have already sketched this phenomenon out for you [Figure 2c]. We were able to say that if we have the prism here and the beam of light here, the light somehow passes through the prism and is diverted upward. And we also said that it wasn't just being diverted. A diversion would take place if a transparent object with parallel surfaces were placed in the path of the light.[1] But a prism with surfaces that meet is placed in the path of the light, and we get a darkening of the light in its passage through the prism. Thus, at the moment we shoot the light through the prism we are dealing with two different things—the simple, bright light that is radiating forth and the darkness that has been placed in its path. However, this darkness, we said, is placed in the path of the light in such a manner that, while the light is diverted for the most part upward, what emerges in the form of darkness, by radiating upward, will have its rays in the same direction as the diversion. In other words, darkness radiates into the diverted light. Darkness lives, so to speak, in the diverted light. In this way the blues and violets arise. However, the darkness also radiates downward. While the beam of light is diverted this way [upward], here the darkness radiates downward, but it is acting in opposition to the diverted light and is no match for it. Thus, we can say that the diverted bright light drowns out the darkness, and we get yellows or reddish-yellow colors.

If we take a sufficiently narrow beam of light, then, by looking in the direction of this beam, we can look with our eyes right through the prism. Rather than looking from the outside at the image thrown upon the screen, we can put our eye in the place of this image. If we then look through the prism, the opening that creates the beam of light will appear displaced to us [Figure 4c]. If we stick with the facts, once again we have here the phenomenon right before us: If I look in this direction, what would otherwise come directly to me is displaced downward because of the prism. Besides, I see it in color. Everywhere you see it in color.

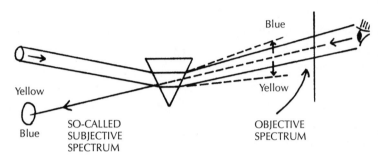

Figure 4c

What are you actually seeing? If you can imagine what you see here, and state what you are seeing purely in the context of what we have just established, then what you are actually seeing will immediately make sense to you, even in the details. All you have to do is stick with what you see. Isn't it true that, if you look at the beam of light this way, you see something bright, because the beam of light is coming toward you, but you see the brightness through darkness, through the blue color—brightness through darkness. Therefore, here you have to see yellow or reddish yellow. Isn't the fact that blue emerges here clear proof that you have something darkened up here? Down here, the red color proves the same thing—that you have something illuminated. As I already explained—the brightness drowns out the darkness. So, by looking here you see the beam of light, however bright it may be, through something illuminated. Compared with the illuminated object, the beam is something dark. So you are seeing a dark object through a bright object, and you have to see it as blue or purple at the bottom. You simply have to state the phenomenon, and then you already have what you can see. What presents itself to the eye is what else you see here—the blue that you are looking through. Thus the brightness appears reddish.

On the lower edge you have an illuminated area. However bright the beam of light may be, you see it through an illuminated object, so you are seeing a darker object through an illuminated object, and it appears blue. That is what it comes down to—polar opposites.

If we want to be scholarly, we can call the first experiment, on the screen, the objective spectrum. The second one—what we see when we look through the prism—we can call the subjective spectrum. The subjective spectrum appears as the inversion of the objective spectrum. If we say it like that, we are speaking like true scholars.

People have speculated a great deal about these phenomena, particularly in the modern era. Not only have they observed the phenomena and described them clearly, as we have tried to do now, but they have also speculated about these things, and the most extreme speculation was attempted when the famous Newton[2] thought about light, because the color spectrum presented itself to him first. Of course, Newton made it relatively easy for himself with his so-called explanation, which is all it's ever been. He said that if we have a prism and let white light in, the light already contains the colors; the prism lures them out, and then they march out in order. We simply split up the white light. Then Newton had the idea that a particular substance corresponds to a type of color, so that seven color substances are contained in the whole thing. For him, the passing of light through the prism is a sort of chemical dissection of light into seven individual substances. He even had ideas about which substances emit larger particles—little spheres—and which emit smaller ones. According to this idea, the sun sends us light, we let the light in through the circular opening, and it strikes here [prism] as a beam of light. However, this light consists entirely of little particles, tiny little bodies, which hit here, are diverted from their direction, and then bombard the screen. The little cannonballs strike here [prism]. The little ones fly up, the big ones down. The little ones are violet and the big ones are red, don't you see? Thus do the big ones separate themselves from the little ones.

Other physicists, Huygens[3] and Young[4] and others, very soon cast doubt on this view that a substance or various substances are flying through the world. Finally they arrived at the conclusion that it just can't work that way. These tiny spheres being emitted from someplace and simply driven or not driven through a medium, then either arriving at a screen and producing an image or reaching the eye and evoking in us the appearance of red and so forth—it just can't work that way. And people were finally driven to prove this. This whole way of looking at things was upset particularly by an experiment carried out by Fresnel,[5] which had already been prepared by the Jesuit Grimaldi[6] and also by others.

Fresnel's experiments are extraordinarily interesting. We have to get a clear understanding of what actually happens in the way Fresnel set up his experiments. I ask you to pay very close attention to the facts because it's a matter of studying a phenomenon exactly. Let's assume that I have two mirrors and a source of light here. In other words, I shine light from a flame at this point, so that, by setting up a screen here, I get images from this mirror and images from the other mirror [Figure 4d]. Let's also assume—I'll draw it in cross section— that the two mirrors are very slightly angled toward each other. Here I have a light source—I'll call it L—and a screen. Thus the light is reflected by striking here [mirror], so that I am able to illuminate the screen with the reflected light. If I allow the light to strike here, then with the mirror I can illuminate the screen here, so that it is brighter here in the middle than in the surrounding area. Now, however, I have a second mirror here, which reflects the light somewhat differently, and a part of what is directed at the screen from my beam of light here below will fall on the upper image. Because of the angle, more or less, the light that the upper mirror reflects is projected onto the screen, as well as that which is reflected by the lower mirror. It is as if the screen were being illuminated from two different places.

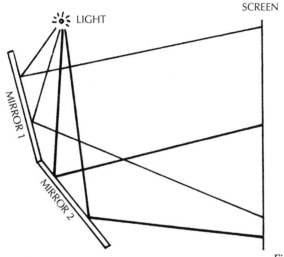

Figure 4d

Let's say that a physicist who thinks in Newtonian terms sees this. He would say to himself: There's the light source. First, it bombards the first mirror, which hurls its little spheres this way. They bounce off, arrive at the screen, and light it up. But the little spheres also bounce off the lower mirror. A lot of little spheres arrive from there. If there are two mirrors, it must be much brighter than if there were just the one mirror. If I arrange things in such a way that I take away the second mirror, then the screen would have to be illuminated less by the projected light than when I have two mirrors. Mind you, a really awkward thought could occur to this physicist, for these particles, these little bodies, have to go this way just as the others are coming down. Why the ones that are coming down don't bump into them at all and knock them away is extraordinarily difficult to understand. In our physics books you can find really very lovely stories about wave theory. However, while things are very nicely calculated, you always have to think that they never calculated how such a wave zips through another one. It always goes completely unnoticed. Let's try to understand just for once what really happens here.

Certainly, the light falls down here, is projected over here, falls onto the second mirror too, and is projected over here. Thus the light goes to the mirror and is projected over here—that's always the way of light. But what actually happens? Now let's assume we have a stream of light like this one, and it's projected across here. But now here comes the other stream of light and collides with it. The phenomenon can't be denied: they disturb each other mutually. That one wants to zip through there; the other one gets in its way. The consequence of this is that if that one wants to zip through there, it first extinguishes the light coming from there. Because of this, however, we don't get brightness here [screen] at all. Instead, in actuality darkness is reflected across here, so that we get darkness here [Figure 4e]. Now, however, all of this is not at rest; it's in continual movement. What has been disturbed here continues. It's as if a hole has arisen in the light. The hole appears dark, but because of it the next body of light will pass through all the more easily, and next to the darkness you will have a spot that is lighter. This continues, and once again a little beam of light from above collides with a stream of brightness and extinguishes it, calling forth darkness once again. We are dealing with a continual lattice; the light coming from above can always get through, and, by extinguishing the brightness, brings darkness once again, which itself continues, however. Thus here we have to get alternating brightness and darkness because the upper light passes through the lower light and makes a lattice.

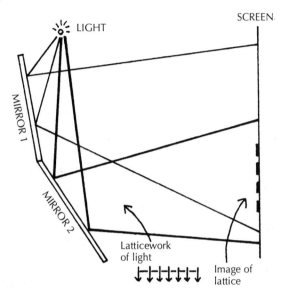

Figure 4e

I have asked you to consider this because you have to look into how a lattice emerges. You have alternating bright spots and dark spots because light is zipping into light. When light zips into light, then the light is simply cancelled. The light is transformed into darkness. We have to explain the emergence of such a lattice of light based on the arrangement we have made with these mirrors. The speed of light, indeed, any differences in the speed of light that occur here, are of no great significance. What I want to show is that what happens here inside the light itself, with the aid of the apparatus, is that the lattice is reflected here [on the screen]: light, dark, light, dark.

But Fresnel thought that, if light is the emission of particles, then it is self-evident that when more particles are emitted, it has to become brighter—otherwise one particle must
have eaten up the other. Thus the fact that brightness and
darkness alternate can't be explained according to the emission theory alone. We have just seen how it can be explained,
but you see once again that it didn't occur to the physicists to
take the phenomenon for what it actually must be. Instead
they attempted an explanation in the spirit of materialism by
connecting it to certain other phenomena. The bombardment
of little spheres of matter no longer worked. Therefore they
said, "Let's assume that light isn't the emission of fine substances, but only a movement in a fine substance, in the
ether—movement in the ether." And first they had the idea
(Euler, for example)[7] that light is propagated in this ether
more or less as sound is in the air. If I produce a sound, of
course it's propagated through the air, but in such a way that
at first the air in the area surrounding where the sound is produced is squeezed together, thereby creating compressed air.
This compressed air presses in turn on its surroundings. It
expands, sporadically causing a rarefied layer of air directly in
its vicinity. Through this compression and expansion, which
are called waves, we imagine the transmission of sound. And
thus they assumed that waves of this kind are produced in the
ether. However, the idea didn't fit in with certain phenomena,
so they said that light was certainly a wave movement, but
that it doesn't vibrate the way sound does. In the case of
sound, there is compression here, then expansion, and it
progresses in the same way. Those are longitudinal waves.
Thus expansion follows compression, and a body moves
within it back and forth in the direction of propagation. With
light, they couldn't imagine this in the same way. There it
must be the case that, when light is propagated, the ether

particles move vertically to the direction of propagation. Thus, when what we call a ray of light zips through the air (such a ray zips along at a speed of 300,000 kilometers a second), the little particles always vibrate vertically to the direction in which the light is propagated. When these vibrations reach our eye, we perceive them.

If we apply that to Fresnel's experiment, then the movement of light is actually a vibration perpendicular to the direction in which the light is propagated. This ray going to the lower mirror would vibrate this way, continue like this, and strike here. Now, as I said, they overlook the fact that these series of waves pass through each other. According to physicists who think like this, the waves don't interfere with each other. But here [at the screen] they interfere with each other right away, or they can also reinforce each other. For what should happen here now? Isn't it true that when this series of waves arrives here, it can be the case that one tiny particle vibrating vertically vibrates downward at the same time that another vibrates upward? Then they cancel each other, and darkness would have to result. If, however, one particle here is vibrating downward just as the other is vibrating downward, or vibrating upward when the other is vibrating upward, then brightness would have to result. Thus, they explain on the basis of vibrations of the smallest particles the same thing that we explained based on the light itself. The so-called wave theory explains the alternating bright and dark spots here by saying that light is a vibration of the ether: if the smallest particles vibrate such that they reinforce each other, a brighter spot results; if they vibrate in opposite directions, a darker spot results.

Now you have only to consider the difference between the pure perception of the phenomena—remaining within the phenomena, and investigating and describing them—and simply making something up about the phenomena. The movement of the ether is after all a pure invention. Of course, we can make calculations about something like this, which we have made up, but the fact that we can make calculations about it is no proof that the thing is there. The purely kinematic is something purely imaginary, and calculations are also imaginary. You can see that, according to the tenets of our way of thinking, we have to rely on explaining phenomena in such a way that they themselves yield their own explanation, that they hold their explanation within themselves, and—I can't emphasize this too much—that everything that is mere musing has to be rejected. You can explain anything by adding something that nobody knows anything about. These waves, for example, could be there, naturally, and it could be that, if one vibrates downward and the other upward, they cancel each other then—but we made them up. However, what is there without a doubt is this lattice, and we see it faithfully reflected here. You do have to look at the light if you want to arrive at an explanation that isn't counterfeited.

Now I told you that if one light passes through another or comes into any kind of contact with it at all, then under certain circumstances one light acts to darken the other light, to extinguish it, just as the prism itself acted to darken the light. That becomes especially clear if we do the following experiment. I'll try to sketch out what it's about. Let's assume that we have what I showed you yesterday: a spectrum from violet to red. We actually have such a spectrum, produced, in fact, directly by the sun. Instead of producing such a spectrum by allowing the sun to shine through such an opening, we could also produce it by placing a solid body here that we cause to glow [Figure 4f]. Then, as it becomes white-hot, we gradually have the possibility of getting such a spectrum. It is of no consequence whether we have a sunlight spectrum or one that comes from a white-hot body.

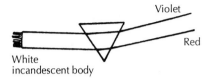

Figure 4f

Now, however, we can also produce a spectrum in a somewhat modified way. Let's say we have a prism here and we have a sodium flame here, in other words a volatile metal: sodium. The sodium turns into gas, which burns and volatilizes, and we produce a spectrum from the volatilizing sodium [Figure 4g]. Then something quite peculiar happens. If we produce the spectrum not from the sun or from a glowing solid, but from a glowing gas, then a single part of the spectrum is very strongly pronounced. In fact, sodium light tends especially to yellow. Here we have red, orange, and yellow. The yellow part is particularly strongly pronounced in sodium. The rest of the spectrum is atrophied, hardly even present, in the metal sodium. Therefore, we apparently get a narrow yellow strip; we call it a yellow band. This happens because it is part of a whole spectrum; the rest of the spectrum is just atrophied. Thus, with all different kinds of bodies we can find such spectra that aren't really spectra at all, just shining bands. From this you can conclude, conversely, that if you don't know what is actually in the flame and you create a yellow spectrum with it, then there must be sodium in the flame. You can recognize which metal you are dealing with.

Figure 4g

An odd thing happens, however, if you combine these two experiments [Figure 4h]. We produce the beam of light here and the spectrum here, and at the same time we put the sodium flame in so that the glowing sodium unites with the beam of light. What happens there is quite similar to what I showed you a little while ago with Fresnel's experiment. We could expect that the yellow here would appear especially strong because there is already yellow in the beam, and then the yellow from the sodium is added to it. But that's not the case. Instead, the yellow from the sodium extinguishes the other yellow, and a dark spot is created. Thus, where we would expect a brighter area to emerge, a dark spot emerges! How could that be? It depends solely on the force that is generated. Let's assume that the sodium light created here was so selfless that it simply allowed the related yellow light to pass through it. Then it would have to extinguish itself completely. It doesn't do that, though. Instead it blocks the way at exactly the point where the yellow should cross. It is there. Although it is yellow, it doesn't act to strengthen. Instead, it acts to extinguish, because as a force it simply blocks the way, regardless of whether the thing that comes into its path is something else or not. That's of no importance. The yellow part of the spectrum is extinguished, and a black spot is created.

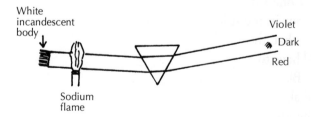

Figure 4h

From this you can see that once again you only need to consider what is there. The flood of light itself offers you the explanation. That's exactly what I would like to point out to you. You see, the physicist who explains things in the spirit of Newton would naturally have to say that if you have something white here, a strip of white for example, and look at this shining strip, then it appears to you that you're getting a spectrum: red, orange, yellow, green, blue, dark blue, violet [Figure 4i].

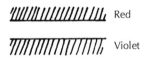

Red

Violet

Figure 4i

Now you see, Goethe would say, "Sure, at a pinch that's all right. If nature really is such that it put light together, then we could assume of course that this light is really dissected into its parts by the prism. Wonderful. But the very same people who say that light consists of its seven parts assert at the same time that darkness is nothing at all but the absence of light. Fine, but if I leave a strip of black here between the white and look through the prism, then I also get a rainbow, only with its colors in a different arrangement. Now it's violet in the middle and becomes bluish green[8] toward one side. Here I get a band that is arranged differently, but in the spirit of the dissection theory I would have to say that the black could be broken down too. Thus I would have to admit that darkness isn't merely the absence of light. Black would have to be divisible too. However, it would also have to consist of seven colors." That's what made Goethe lose his faith[9]—he also saw the black strip in seven colors, only in a different arrangement.

This is what forces us once again simply to take the phenomena as they are. Now, we'll have to see to it that tomorrow at eleven-thirty we're able to demonstrate what I unfortunately could only explain theoretically today.

Fifth Lecture

AS WELL AS WE CAN with our limited means, we are going to begin today by showing you the experiment we spoke about yesterday. You undoubtedly remember that I said that when a glowing solid emits light and we send this light through a prism, we get a spectrum, a light image similar to that of the sun. When we have a glowing gas that emits light, we also get a light image, but it shows actual light lines or light bands in only one place (or, for various substances, in several places). The rest of the spectrum is atrophied. By carrying out exact experiments, we would see that actually a complete spectrum is present for everything that glows, in other words a spectrum that ranges from red to violet. If, for example, we produce a spectrum with glowing sodium gas, then we simply have a very, very weak spectrum and in one spot a stronger yellow line, which by contrast mutes everything else. We therefore say that sodium yields nothing but this yellow line.

Now there is an odd fact, which was known by many earlier and then reconfirmed by the experiment of Bunsen and Kirchhoff[1] in 1859: If we have a light source that produces a continuous spectrum and, acting simultaneously so to speak, a light source that produces something like the sodium line, then this sodium line acts simply like an opaque body; it opposes the very color quality that would be at that spot—in this case yellow—and extinguishes it, so that instead of the yellow we have a black line there. In other words, if we stick to the facts, what we can say is that for the yellow in the spectrum another yellow, which has to be at least as strong as the yellow generated at this spot, acts like an opaque body. As you will see, there will be a basis for understanding this in the elements we are putting together. First, we must stick to what is factual.

Now we will show, as well as we can, that this black line is really in the spectrum when we add in the glowing sodium. We aren't able to do the experiment here in such a way that we capture the spectrum, so instead we'll do it a way that lets us study the spectrum by looking at it with our eyes. That way we can also see the spectrum, but instead of being displaced upward, it will be reversed, displaced downward, and the colors will be reversed. Remember that we spoke about why the colors appear like this when I simply look through the prism. We'll produce the beam of light with this apparatus here, let it pass through here, and look at the broken beam of light here. Thus at the same time we're looking at that, we'll see the black sodium line. I hope you'll be able to see it, but you'll have to approach and look in perfect military order—which shouldn't be too difficult in Germany right now. [The experiment is shown to everyone individually.]

Now we still want to use the short time remaining to us, so we'll have to move on to a discussion of the relationship of the colors to the so-called bodies. As an approach to the problem of finding this relationship, I want to show you the following. You see the complete spectrum captured on the screen. In the path of the light beam I now place a small trough containing carbon disulfide in which some iodine is dissolved, and ask you to observe the resulting change in the spectrum. You see that you have a clear spectrum here, and when I place the solution of iodine in carbon disulfide in the path of the light beam, it extinguishes the light completely. Now you see the spectrum clearly separated into two parts because the middle part has been extinguished. Thus you see only violet on one side and yellowish red on the other. In this way, because I let the light pass through the solution of iodine in carbon disulfide, you see the complete spectrum separated into two parts, and you see only the two poles.

I have really lost a lot of time and will only be able now to talk about some principles. Isn't it so that the key question concerning the relationship of the colors to the bodies we see about us (and all bodies are colored in a certain way) must be to explain how it is that these bodies appear colored—in other words how they have a certain relationship of their own to light, how they develop a relationship to light through their material being? One body appears red, another blue, etc. Naturally, the simplest way to manage is by saying that when colorless sunlight—which for the physicist means a collection of all colors—falls on a body that then appears red, that stems from the fact that the body swallows all the other colors and reflects only red. It is also easy to explain how a body is blue: it simply swallows all other colors and reflects only blue. For us it a matter of completely excluding such a speculative explanatory principle and of approaching the somewhat complicated fact that we see the so-called colored bodies by way of the facts—by lining up fact upon fact in order to capture something that presents itself as the most complicated of phenomena.

The following will lead us onto the path. We remember that in the seventeenth century, when people were still doing a lot of alchemy, they spoke of the so-called phosphorous substances, the light-bearers. At that time, by "phosphorus" they meant the following. A shoemaker in Bologna,[2] to take an example, was experimenting alchemically with a kind of barite, the so-called Bologna stone. He exposed it to light, and the remarkable phenomenon occurred that whenever he exposed the stone to light, it shone with a particular color for a time afterward. Thus, the Bologna stone had developed a relationship to light, which it expressed by continuing to shine even after the light had been taken away. Therefore such stones, which were being investigated in various ways after this fashion, were called "phosphorus." So if you encounter the expression "phosphorus" in the literature of this time, you shouldn't take it to mean what we mean by it today, but rather phosphorescent bodies, light-bearers, phosphores. Now, however, this phenomenon of the afterglow, of phosphorescence, isn't actually the basic one; a different phenomenon is the basic one.

If you take ordinary oil and look through it toward something that is shining, you will see the oil as faint yellow. If, however, you place yourself so that you allow the light to pass through the oil and look at it from behind, the oil will appear to glow with a bluish light, but only as long as light is shining on it. You can carry out this experiment with various other bodies. It gets particularly interesting when you dissolve chlorophyll. If you look through such a solution into the light, it appears green. But if you place yourself behind it, so to speak, with the solution here and the light passing through here, so that you now see the place where the light passes through from behind, then the chlorophyll shines back with a reddish color—red, just as the oil shines blue [Figure 5a].

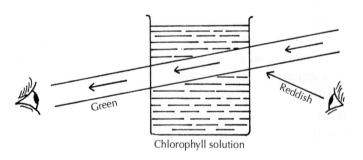

Chlorophyll solution

Figure 5a

There are bodies of the most various kinds that shine differently when they themselves send light back, so to speak, than when the light passes through them as through a transparent body. That is, they show that they have entered into a relationship with the light, which is transformed by their own nature. If we look at the chlorophyll from behind, we see what the light has accomplished in the chlorophyll, so to speak, the relationship between the light and the chlorophyll. This phenomenon of a body shining with a light while being illuminated by another light is called fluorescence. And we can ask, "But what is phosphorescence?" It's just a kind of fluorescence that lasts. When chlorophyll, for example, appears reddish as long as light is acting on it, that is fluorescence. With phosphorescence, we can take away the light, and barite, for example, continues to shine a little. In other words, it retains the characteristic of giving off colored light, while chlorophyll does not. So you have two stages: One is fluorescence: we cause a body to be colored as long as we shine a light on it. The second stage is phosphorescence: we cause a body to continue to be colored for a certain time afterward. And then there is a third stage: the body appears permanently colored by means of something that light does with it. Fluorescence, phosphorescence, a body's state of being colored—here we've lined the phenomena up next to each other, so to speak. Now it's only a matter of approaching the phenomena with our ideas in an appropriate way.

To do that, you will need to take in a certain idea that we will be working on together with all of this. Now I'll ask you once more to think absolutely only about what I'm presenting to you and to think as exactly and precisely as possible, and I'll remind you of the formula for velocity v. As you know, any velocity, however fast, is expressed, by dividing the distance *d* by the time *t*, so that the formula reads

$$v = \frac{d}{t}$$

Now the opinion persists that somewhere in nature there is a distance d in space that has been traveled and a time t during which the distance in space was traveled, and that we then divide the real distance in space by the real time to get the velocity, which we actually regard as something that is not quite real. Rather we regard it more as a function, as something we arrive at as the result of calculations. That is not how it is in nature. Of these three quantities—velocity, space, and time— velocity is actually the only real one. Velocity is the one that is outside us; we arrive at the others, d and t, only by dividing, by splitting, so to speak, the unified v into two abstract things, which we create on the basis of the existing velocity. We proceed more or less as follows: We see a so-called body fly with a certain velocity through space. The sole reality is that it has velocity. However, instead of contemplating the totality of this flying body, we now think in two abstractions. We divide something that is a unity into two abstractions. Because there is a velocity, there is a certain path. We consider that first. Then we consider the time during which this path is covered, isolating space and time by our thinking process. However, this space is only there because the velocity makes it, and the time is no different. Space and time, as related to this real thing, which we designate with v, are not realities, but abstractions, which we just fashion from the velocity. And we will not be able to cope with outer reality until we realize that we in our thinking process have created this duality of space and time—that velocity is the only real thing outside of us, if you like, and we only created space and time from the two abstractions into which velocity can be divided.

We can detach ourselves from velocity, but we can't detach ourselves from space and time: they are integral to our perception, to our perceptual activity. We are one with space and time. Think about that. We are not one with the velocity outside, but with space and time. Indeed we shouldn't so quick to ascribe to external bodies that with which we ourselves are one. Rather we should use the fact that we are one with space and time in an appropriate way to arrive at a conception of external bodies. We should say, "Because of space and time, with which we are intimately connected, we learn to recognize velocity." However, we shouldn't also say that a body requires time. Rather only: "A body has velocity. By means of space and time we measure velocity. Space and time are our instruments, and they are bound to us; that is the important thing." Here in space and time, you see once again the so-called subjective clearly differentiated from the so-called objective, which is velocity. It will be very good if you understand this very, very clearly, for then it will become apparent to you that v isn't merely the quotient of d and t. Rather what I express here with the number has an inherent reality on its own terms, whose essence consists in having a velocity. What I have shown you here for space and time—that they are not separable from us at all, that we may not separate ourselves from them—is also true of something else.

These days there is still a lot of "Königsbergism" in peo-ple—I mean Kantianism. This Königsbergism has to be gotten rid of completely because somebody could believe that just now I had spoken in the spirit of Königsbergism. In that case we would say, "Space and time are in us." But I'm not saying that space and time are in us. Rather, in perceiving the objec-tive, velocity, we use space and time for perception. Space and time are simultaneously inside us and outside us, but we form a bond with space and time, whereas we don't form a bond with velocity. The latter roars right past us. In other words, it is something essentially different from Kantian Königsbergism.

What I have said about space and time is also true of another thing. Just as we are connected to objectivity by space and time, but have to search first for velocity, we are in the same element with the so-called bodies, in that we see them by means of light. We may no more speak of the objectivity of light than we may speak of the objectivity of space and time. We float in space and time, just as bodies float in it at a certain velocity. We float in light, as bodies float in light. Light is an element we share in common with what is outside of us in the form of so-called bodies. You can imagine it like this: If you have gradually illuminated the darkness with light, then space fills up with something—let's call it x, if you like—something that both you and that thing outside you are inside of, a common element, in which you and the elements are floating. Now we have to wonder how we do that—how we float in the light there. We can't float in it with our so-called body, but we do float in it with our etheric body. We won't be able to understand light unless we get on to realities. We float in the light with our etheric body—if you like, you can say, in the light ether; that's not so important. So—we float in the light with our etheric body.

Now in the course of our time here we have seen how colors emerge in all different ways in and under light. Colors also emerge in the so-called bodies or persist in them. We see ghostly colors, so to speak, which emerge and disappear in light. If I just project a spectrum, it is like ghosts—it flits about in space, so to speak. In and under light we see colors of this sort. How can that be? We float in light with our etheric body. How are we related to the colors that flit about? The only possibility is that we are inside them with our astral body. We are joined to the colors with our astral body. There's no alternative but to realize that wherever you see colors, you are joined to the colors with your astrality. In order to achieve real insight there's no alternative but to say to yourself that although light actually remains invisible, we are floating in it. Just as we shouldn't call space and time objective entities because we float in them with things, we should also regard light as our common element. We should regard the colors, however, as something that can appear only because we enter with our astral body into a relationship with what the light is doing there.

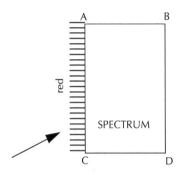

Let's assume, however, that somehow you have produced some kind of color phenomenon, some kind of spectrum, in this space here, A-B-C-D, but a phenomenon that happens only in the light [Figure 5b]. Here you will have to go back to an astral relationship with the light. However, you could have colored this here as the surface so that A-C as a body appears red to you, for example. We say, "A-C is red." Then you look at the surface of the body and at first have the rough idea that under the surface of the body it is red through and through. You see—that is something different. There you also have an astral relationship, but you are separated by the surface of the body from the astral relationship you enter into with the color. Try to conceive of that! In the light you see colors, colors of the spectrum. There you have astral relations of a direct nature— nothing comes between you and these colors. You see the colors of the body; something comes between them and your astral body and yet you enter into astral relations with the colors of the body. I'd like you to take these things to heart and think them through clearly because they are important fundamental concepts, which we will be working on. And only by doing that will we develop fundamental concepts for a real study of physics.

In closing, I would just like to mention one more thing. I am not trying to talk to you about what you could easily get for yourself by buying the first textbook that comes along. I also don't want to tell you about what you can read if you read Goethe's *Theory of Color*. Rather I want to tell you about what you can't find in either, but which will enable you to nourish yourself intellectually from both in a suitable way. Even if we aren't true believers in physics, we don't need to become true believers in Goethe. Goethe died in 1832, and we don't confess to an 1832 Goetheanism, but rather to one of the year 1919—in other words, to a Goetheanism that has continued its education. I would particularly like you to think over what I said to you today about the astral relationship.

Sixth Lecture

TODAY I WOULD LIKE to continue my examination for you of the principal ideas that we began with the day before yesterday. If we start out from the knowledge we have gained with light, we will be able to observe and understand further the phenomena that can be revealed in the other natural phenomena we still want to study. So today I will concentrate on the principles and postpone the experimental work until tomorrow since we still have to figure out just what methods we want to use. It's really a matter of precisely following through to their conclusion what is present in the natural phenomena. And light gives us the most clues for pursuing that course.

Historically it so happened that people began to study light phenomena relatively late. In general the whole way of thinking about physical phenomena   as it is done in our schools today hardly goes back further than the sixteenth century. Before the sixteenth century, the way of thinking about physical phenomena was radically different. Today, however, this way of thinking has been taken up so strongly in school that it becomes extraordinarily difficult for anyone who has gone through a certain kind of schooling in physics to return to the purely factual. First, you have to get used to feeling—please don't take the expression merely in its trivial sense—to experiencing the purely factual. You actually have to get used to it. For that reason, I want to start out with a specific case, comparing the customary didactic way of thinking with what we can actually gain through a proper pursuit of the facts.

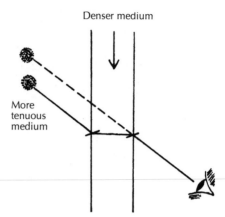

Figure 6a

Let's assume that you have the cross section of a plate of glass here. Through this plate of glass you see a shining object. I will sketch the thing, but instead of the shining object, let's say I simply draw a shining circle here [Figure 6a].[1] Now imagine yourself back on the school bench once more and recall what you actually learned from this vantage point about visual observation. You were told that rays emanated from this shining object (we're interested in a certain line of sight for the eye); in other words, in the direction of this ray the light penetrates, as they say, from a thinner medium into a denser medium. If we simply look through the plate of glass and compare what happens with what is really there, we can perceive that the shining object is displaced and appears at a different spot than when we look at it without the plate of glass. This is said to stem from the fact that the light is bent. That's what they say when the light enters a denser medium from a thinner medium. Then, in order to figure out the direction, we have to draw a so-called angle of incidence. If the light continued on its way without being hindered by a denser medium of this kind, it would go in this direction. However, the light *is* bent, as they say, and in fact it is bent toward this perpendicular of incidence, toward this perpendicular line that we drop at the point of incidence. On the other hand, if you follow how we look at the ray through the denser medium, we have to set up a perpendicular of incidence where it exits as well. At this point, the ray, if it were to continue on its path, would go this way, but it is bent once again, in this case away from the perpendicular of incidence, in fact, and just strongly enough that its direction is now parallel to its former one. When the eye sees that, it lengthens the last direction for itself and shifts the shining object a certain distance higher up. Thus, if we look through like this, we have to assume that the light falls here and is bent twice, once toward the perpendicular of incidence and once away from the perpendicular of incidence. Because the eye has this inner ability (or a soul or some demon, however you want to say it), the light is shifted in space, in fact to another spot in space than where it would appear if we didn't see it through a refractive medium, as they call it.

However, it is important to keep the following in mind. You see, if we look through the same denser medium and try to make a slight distinction between, say, a brighter spot and a somewhat darker spot,[2] we find that not just this brighter part is displaced upward, but the somewhat darker part is as well [Figure 6b]. We will see the whole complex you see here displaced upward. I would like you to take that into consideration. Here we see a darker area bordered by a brighter one. We see the darker area displaced upward, and, because it has one brighter end, we see that displaced upward too. So if you put up a complex of this kind, a darker area and a brighter area, you have to say that the brighter part is displaced simply as the upper border. If you abstract a bright spot, then people often speak as if only this bright spot were displaced. But that's an absurdity. For even if I look at this bright spot here, it is not true that it is the only thing that is displaced. Instead, in reality this area down here, which I'll call "nothingness," is also displaced upward. Whatever is displaced is never something I can delimit so abstractly.

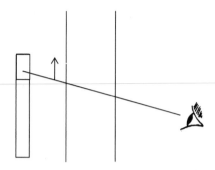

Figure 6b

Thus if I do the experiment that Newton did—if I let in a beam of light, which is diverted by the prism—then it isn't true that only the beam of light is displaced; what borders the beam of light, above and below, is displaced too. I should never speak of any sort of rays of light or the like, but of displaced light images or light spaces. And if somewhere I do want to talk about an isolated light, then I can't talk about it at all in a way that relates this isolated light to something in the theory. Rather I have to talk about it in such a way that my words refer simultaneously to what borders it.

Only if we think in this way can we really feel what is actually happening when we are faced with the origin of color phenomena. Otherwise, simply because of our way of thinking, we get the impression that the colors somehow arise from light—we have already worked out the thought that we are dealing only with light. In reality, we aren't dealing with light, but with something bright, which is bordered on one side or the other by darkness. And, just as this bright area is displaced as light space, the dark area is displaced in the same way. But what is this dark area? What is it actually? You see, this dark area also has to be grasped as something absolutely real. Everything that has come into modern physics since approximately the sixteenth century was able to come in only because we never observed things with the spirit; we observed things only on the basis of outward sense impressions and then invented all sorts of theories to explain these external sense impressions. You won't be able to deny in any way that, if you look at light, sometimes it's stronger and other times it's weaker. Stronger and weaker light do exist. It's a matter of understanding how this light, which can be stronger or weaker, is related to the darkness.

The ordinary physicist of today thinks there is stronger and weaker light, every possible degree of light in terms of strength, but only one single darkness, which is simply there when light isn't there. Thus there is only one kind of "black." There is no more just one kind of darkness than there is just one kind of brightness; to say that there is only one kind of darkness is as one-sided as if you were to say, "I know four people. One has assets of five hundred marks and the second assets of a thousand marks. In other words, one has greater assets than the other does. The third, however, has a debt of five hundred marks and the fourth a debt of one thousand marks. But why should I worry any more about this difference? In the end, it's the same thing. Both have debts. I want to distinguish between degrees of wealth, but I don't want to distinguish at all between degrees of debt because debt is debt." In this case the matter attracts our attention because the effect of five hundred marks of debt is less than the effect of a thousand marks of debt. With darkness we don't act this way: light has different degrees of brightness; darkness is darkness.

The fact that we haven't progressed to a qualitative way of thinking is what hinders us so much in finding the bridge between the soul-spirit realm on the one hand and the physical on the other. When a space is filled with light, it is filled with light of a certain strength; when a space is filled with darkness, it is filled with darkness of a certain strength. We have to progress from merely abstract space to space that is not abstract, but that is in some way positively filled with light, negatively filled with darkness. Thus we can face the light-filled space and can call it qualitatively positive; we can face the space filled with darkness and find it qualitatively negative with reference to the light. Both, however, can be addressed in terms of a particular degree of intensity, a particular strength. Now, however, you wonder, "Yes, but how does this positive state of light-filled space distinguish itself for our powers of perception from the negative state of filled-up space?" For this positive state of filled-up space, we need only to recall how it is when we wake up, surrounded by light, and unite our subjective experience with what floods around us as light. We need only to compare this feeling with what we feel when we are surrounded by darkness, and we will find that—I'd like you to understand this very precisely—for our feelings there is a difference between abandoning ourselves to light-filled space and abandoning ourselves to space that is filled with darkness. We can approach these things only by comparison.

You see, we can compare that feeling we have when we encounter the light-filled space to a kind of breathing in of the light by our soul nature. Indeed we feel enriched when we are in a light-filled space. We breathe the light in. What is it like with darkness? That feeling is the complete opposite. Darkness drains us; it soaks us up; we have to abandon ourselves to it; we have to give up something to it. Thus we can say that light has an imparting effect on us and darkness has a draining effect on us. And we have to distinguish between bright and dark colors in this way too. The brighter colors have something that comes toward us, that imparts itself to us; the dark colors have something that drains us, that we must give ourselves up to. Then we come to the point where we say to ourselves that something in the outer world is imparting itself to us when light acts upon us and that something is taken away from us, drawn out of us, when darkness acts upon us.

As I've already pointed out to you in these lectures, our consciousness is in a certain respect drained away when we fall asleep. Consciousness ceases at that moment. When we approach the darker colors, blue and violet, from the brighter ones, there is a very similar phenomenon of the cessation of consciousness. And if you'll recall what I told you recently about the relationship of our soul nature to mass—the falling asleep into mass, the draining away of consciousness by mass—then you will feel something similar in the draining away of consciousness by darkness. You will discover the inner relationship between the darkness that can fill   space and that other way in which space can be filled, which we call matter, and which manifests itself as mass. In other words, we have to search for the path leading from the phenomena of light across to the phenomena of material existence. And we have already blazed the path for ourselves by searching out the fleeting phenomena, as it were, of phosphorescence and fluorescence and then the stable light phenomena. In the case of the stable light phenomena we have permanent colors. We can't look at these things individually. First we will want to describe the whole complex of things.

Now it's a matter of understanding the following. If you are in light-filled space, you merge with this light-filled space in a certain way. We can say that something in us floats out into this light-filled space and merges with it. However, you have only to reflect a little bit on the facts to discover a big difference between this oneness with the immediate light-flooded surroundings and the sort of oneness that we have as human beings with, for example, the temperature of the surroundings. We take part in the conditions of warmth of our surroundings—we take part in it by feeling something like a polarity of the temperature: warmth and cold. However, we can't avoid perceiving a difference between our feeling of self in the surrounding conditions of warmth and our feeling of self in the surrounding conditions of light. Not only has this distinction been lost to modern physics since the sixteenth century, but we have also striven to blur any differences of this kind. Anyone who really looks at this difference between living through the conditions of warmth and living through the conditions of light, which in the real world is simply a given, can't avoid making the distinction that we are engaged with our physical body in the conditions of warmth, but with our etheric body in the conditions of light. However, what we perceive with our etheric body is confused with what we perceive with our physical body.

This confusion has been a malady for the modern study of physics since the sixteenth century, and because of it everything has gradually been blurred. Especially since physics gradually came under the Newtonian influence, which is still effective today, we have forgotten how to express factual situations directly. Individual people have of course attempted to refer to the unmediated facts—Goethe on a large scale and people like Kirchhoff, for example,[3] in a more theoretical way. However, on the whole, we have actually forgotten how to direct our attention purely to the facts. Thus, for example, we have interpreted in the spirit of Newton the fact that material bodies near other material bodies fall toward them under certain conditions. We have ascribed this to a force that emanates from one body and is exerted on the other one—gravity. You can cogitate all you want, but you will never be able to count what we understand by the word "gravity" among the facts. When a stone falls to the earth, then the fact of the matter is merely that it approaches the earth. You see it at one location, then at second location, at a third location, and so on. If you say that the earth attracts the stone, you are thinking up something besides the facts of the situation. You are no longer purely expressing the phenomenon. We have grown more and more unaccustomed to expressing the phenomenon purely. Yet that is what is important, because if we don't express the phenomena purely, but proceed to made-up explanations, then we can find all kinds of made-up explanations, which often explain the same thing.

Assume you have two planetary bodies. You could say that these two bodies attract each other mutually. They send some unknown like a force out into space and attract each other mutually [Figure 6c]. However, you don't need to say that these bodies mutually attract each other. You can also say that one body is here, the other body is here, and here are also many other little tiny bodies, little ether particles even, if you like, in between [Figure 6d]. These little ether particles are in motion and bombard both planetary bodies. Here they're bombarding this way, there they're bombarding that way, and the ones in between are flying back and forth and bombarding too. Now the area of attack here is larger than the one inside there. For that reason there is less bombarding inside. The result is that the planetary bodies approach each other. They are pushed against each other because of the difference between the number of impacts here in between and the number of impacts made outside. There have been people who have explained gravity by saying that a force acting at a distance attracts the bodies. And there have been people who said that's nonsense, that it is unthinkable to assume that a force acts at a distance.[4] They therefore assume that space is filled up by the ether, and add in this bombardment so that the masses will be whisked into each other. Besides these two explanations, there are all sorts of others. This is just a classic example of how we don't look at the real phenomenon, but invent all kinds of explanations.

Figure 6c

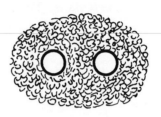

Figure 6d

But what is the real reason for this? This inventing of all kinds of unknown agencies, illusory forms of energy, which do all sorts of things—that spares us something. Of course, this theorizing about impacts was just as much an invention as the theorizing about long-distance forces. But this invention relieves us of an assumption that is frightfully uncomfortable for people today. For, you see, it is always the case that we have to wonder, if there are two mutually independent planetary bodies approaching each other, and they show it is in their nature to approach each other, then, of course, there has to be a basic principle   causing their approach. There has to be a reason for their approach. It is simpler to make up forces than to say that there is yet another way, namely thinking that the planetary bodies are not independent of each other. If, for example, I lay my hand on my forehead, it won't occur to me to say, "My forehead attracts my hand." Instead I will say, "That is an inner act carried out by something that has its basis in the soul and spirit." My hand is simply not independent of my forehead. They are not actually two things—the hand and the forehead. I will only succeed in looking at the situation correctly if I regard myself as a whole. I'm not actually looking at reality if I say, "There's a head, there are two arms with hands on them, there's a torso, and there are two legs." That's not a complete examination. Rather it's a complete examination if I describe the entire unified organism, if I describe the things in such a way that they belong together. That is to say, my task is not merely to describe what I see but to think about the reality of what I see. Just because I see something doesn't yet mean it is real.

Take a cube of rock salt. In a certain sense it is a whole (everything is a whole in a certain sense). It can exist by means of the complex of what is within its six sides. If, however, you look at a rose you have cut, this rose is not a whole because it can't exist by means of the complex of what is in it. The rose can exist only because it's on the rosebush. For that reason the cut rose, although you perceive it just as well as the cube of rock salt, is a true abstraction. It's something that may not be addressed as a reality on its own at all.

Something extraordinarily important follows from this. It follows that, when confronted with each phenomenon, we have to investigate to what extent it is a reality or only something that has been cut out of a whole. If you look at the sun and the moon or the sun and the earth on their own, naturally you might as well make up a force of gravity—a kind of gravitation—just as you might invent a kind of gravitation when my forehead attracts my right hand. But when you look at the sun and the earth and the moon, you're looking at things that aren't whole. Rather they are the limbs of the entire planetary system.

This is the most important thing—that we observe to what degree something is a whole or is cut out of a whole. Innumerable mistaken notions arise because we regard something that is only a partial phenomenon as a whole. However, you see, by looking at partial phenomena and making up forms of energy, we have spared ourselves the trouble of looking at the life of the planetary system. In other words, we have attempted to treat as a whole that which in nature is only a part and then to explain by theories all the effects that emerge.

To sum up, in everything that we encounter in nature, it's important to ask, "What is the whole it belongs to? Or is itself a whole?" In the end we will find whole things only in a certain respect, for even a cube of rock salt is a whole only in a certain respect. Even such a cube can't exist unless there is a certain temperature or other conditions. Actually, in every instance it is necessary for us not to look at nature in such a fragmented way, as is so commonly done.

So, you see, only because we look at nature in such a fragmented way have we gotten into the situation of creating that peculiar figment called universal inorganic inanimate nature. An inorganic inanimate nature doesn't exist, any more than your skeletal system exists without, say, your circulatory system. Just as the skeletal system only crystallizes out of your entire organism, so-called inorganic nature doesn't exist without all of underlying nature, without soul- and spirit-nature. This lifeless nature is the dismembered skeletal system of all of nature, and it is impossible to look at inorganic nature by itself the way we have been looking at it in Newtonian physics since the sixteenth century. Nevertheless the Newtonian approach to physics set out to strip everything down to pure "inorganic nature." But inorganic nature is present only when we ourselves build machines, when we ourselves put something together from the parts of nature. However, that is radically different from the way the so-called inorganic itself exists in nature. The only really inorganic things are our machines, and these in fact only to the extent that we put them together with combinations of natural forces. It's only the condition of being put together that makes them inorganic. Other kinds of inorganic things are only abstractions. Modern physics arose on the basis of this abstraction. It is nothing more than an abstraction that takes for reality what it has abstracted and then tries to explain everything it encounters according to its theoretical assumptions. In reality, you see, we actually can't help but form our concepts, our ideas, with what we are given externally in the world of perception.

Now one phenomenological field has provided us, if I may say so, with an extremely convenient fact. If you strike a bell and near to it you place some device that is light and easily moved, you can demonstrate that the parts of this ringing bell are also vibrating. If you take a reed pipe, you can demonstrate that the air inside the pipe is vibrating, and, on the basis of the movement of the air or of the bell particles, you can notice a connection in terms of the pitch or sound phenomena between the vibrations a body or the air makes and the perception of the tone. In this phenomenological field it's evident that when we hear tones we are dealing with vibrations in the surrounding area. Thus there is a connection—which we will talk about further tomorrow—between sounds and vibrations of the air.

If you are going to proceed abstractly, you can say that we perceive sound by means of the organs of hearing. The vibrations of the air strike the hearing organ, and when they do, we perceive the sound. Then, since the eye is naturally also a sense organ, you can perceive colors with the eye and say that it must be a similar situation. Thus some kind of vibration has to strike the eye. However, it can't be the air—that is quickly ascertained. So it's the ether. Thus we construct the idea, if I may say so, purely by playing with analogies: If air strikes our ears, and we sense a sound, there is a connection between vibrating air and the sensation of sound; if the vibrations of the hypothetical ether collide with our eyes, then, in a similar way, a sensation of light is conveyed. Then we attempt to determine how this so-called ether vibrates through tests like those we have become acquainted with experimentally in these lectures. In other words, we imagine an ether ocean and calculate how things are supposed to happen in it. We calculate something that refers to an entity that we are of course unable to perceive, which we are only able to assume theoretically.

As you've already gathered from the trifles we have gone through experimentally, what takes place within the world of light is extraordinarily complicated, and up through certain periods of the more recent development of physics it has been assumed that a vibrating ether, a fine elastic substance, lies behind, or actually *in* everything that manifests itself as the world of light, as the world of colors. Since we are easily able to determine the laws according to which elastic bodies collide and repel each other, we can calculate what these little goblins in the ether do simply by regarding them as little elastic bodies, by imagining the ether as something that in itself is elastic, so to speak. In this way you can arrive at explanations of the phenomena that we demonstrated here when we created a spectrum. Various kinds of ether vibrations are separated from each other, which then appear to us as the various colors. By certain calculations we can also manage to make comprehensible, based on the elasticity of the ether, that canceling out of the colors, for example of the sodium line, that we demonstrated here the day before yesterday.

In addition to these phenomena, others have come up in recent times. We can devise a light spectrum and extinguish or produce the sodium line in it—whichever you want—and produce the black line; then, besides producing this whole complex, we can also cause an electromagnet to act upon the beam of light in a certain way, and, behold, the electromagnet has an effect on the light phenomenon. For example, the sodium line is canceled at its place, and two others appear, purely due to the effect of electricity, which is always associated with magnetic effects in some way. Thus something that is described to us as electrical forces has an effect on events that we regard as light phenomena, behind which we imagine nothing more than the elastic ether. The fact that we have observed the effect of electricity on this light phenomenon has now led us to assume a relationship between light and electromagnetic phenomena. Thus things have been shaken up a bit in recent times. Before, we could rest on our laurels because we hadn't yet perceived this interaction. Now, however, we have to say to ourselves that they must have something to do with each other. Currently this has led many physicists to see an electromagnetic effect in the transmission of light—that what is passing through space actually consists of electromagnetic rays.

Now think about what has happened here. What has happened is the following. Earlier we assumed that we knew what was behind light phenomena: vibrations, undulations in the elastic ether. Now, because we have become acquainted with the interactions between light and electricity, we've come to the point where we have to look at what actually vibrates as electricity, as radiant electricity—please think about this matter carefully! We want to explain light and color. We trace them back to the vibrating ether. There is something passing through space. Then we believe that we know what light actually is— vibrations of the elastic ether. But now it has become necessary to say that these vibrations of the elastic ether are electromagnetic currents. Now we know what light is even more precisely than before. It consists of electromagnetic currents—only we don't know what these currents are. Thus, we have taken the wonderful path of assuming a hypothesis, of explaining the sensible by means of the supersensible unknown of the undulating ether. Gradually we have been forced once again to trace the supersensible back to the sensible, confessing at the same time, however, that we don't know now what the latter is. It is indeed a most interesting path that we have traveled, from a hypothetical search for an unknown to the explanation of that unknown by means of another unknown.

The physicist Kirchhoff was appalled enough to say that if these more recent phenomena make it necessary not to believe in the ether with its vibrations, then that is no advantage for physics; and Helmholtz,[5] for example, when he became acquainted with these phenomena, said, "Fine, we are not going to get around regarding light as a kind of electromagnetic radiation. Then we're just going to have to trace it back to vibrations of the elastic ether again." The essential point is that we have transposed an honest-to-God wave phenomenon— vibrations of the air when we perceive sounds—by way of pure analogy into a field where the whole assumption is simply a completely hypothetical one.

I had to give you this explanation of principles so that we can swiftly go through the most important questions presented by the phenomena that we still want to look at. In the time that remains to us after having laid this foundation, I intend to discuss sound, temperature, and electromagnetic phenomena, and also how these phenomena reflect on optical phenomena.

Seventh Lecture

TODAY WE BEGIN with an experiment that will tie in once again to our observations on color theory. Of course, as I've said, it's only possible for me to present you with something of an improvised, more or less aphoristic nature in these lectures. For that reason I'll also have to avoid the usual categories you find in physics books. I don't mean to say that it would be better if I could keep to these categories; however, I would like to guide you to a certain insight into the natural sciences, so please regard everything I present before that as a kind of preparation, which isn't done by progressing in a straight line, as is otherwise the custom, but by gathering the phenomena we need and creating a circle, so to speak, then pressing forward to the central point.

As you have seen, when colors appear, we are dealing with an interaction of light and darkness. Now it's a matter of observing as many real phenomena as possible before coming to conclusions about the actual causes of this interaction of light and darkness. So today I would first like to show you the phenomenon of the so-called colored shadows.

Using this rod I'm going to produce shadows from two sources of light, represented by these little candles here, on the screen across from you [Figure 7a]. You see two shadows that have no distinct color. If you take a close look at what is here, you have to say that the shadow you see here on the right side is produced by this light source (left), of course, and that it results from the fact that the light is emitted from this source and blocked by the rod. And that shadow is the one that results when the light from our right-hand light source is blocked. Basically, in other words, we are dealing here only with the creation of certain dark spaces. What lies in shadow is just a dark space. If you look at the surface of the screen outside the two shadow bands, you will realize that it is illuminated by the two light sources. So, in other words, we are dealing here with light.

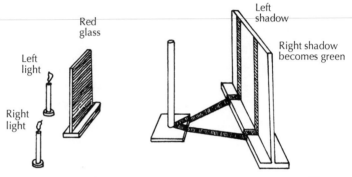

Figure 7a

Now I am going to color one of the lights. I will have it pass through a tinted glass plate so that one of the lights is colored. We know what's going to happen now: one of the lights will be dimmed. But now, you see, because of the dimming of the light, this shadow [right]—which is caused by the rod covering my left light source—turns green. It turns green in the same way that, for example, a white surface turns green when you look directly at a small red surface, then avert your eyes from it and focus them on the white surface. Then what you first saw as red turns green, without anything being there. It's as if you were projecting the green color itself onto the surface. Just as in that case you see the green surface as the afterimage of the red surface you saw before, here you see the shadow of the light source because I am dimming it with red. Thus, what was formerly merely darkness you now see as green. Now watch what happens if I dim the same light source with green! As you see, then the shadow appears red. If I dim the same light source with blue, then, as you can see, the shadow appears orange. If I were to dim the light source with violet, then yellow would result.

Now I'd like you to consider the following—this particular phenomenon is of great significance, so I'll mention it one more time. If you have, say, a red cushion with a white crocheted cover with diamond-shaped openings, and you look first at the red diamonds and then at the white part, you see the same lattice effect of green on the white. Of course, it's not really there, but your eye creates an aftereffect, which produces so-called subjective green images when you focus on the white. Goethe knew of this last phenomenon, and he was also acquainted with the phenomenon of the colored shadows. He said to himself, "I dim this light source, and I get green." And then he describes it as follows: "If I dim the light source here, then the entire white screen is covered with a red light, and I don't actually see the white screen. Instead I see a red light. I see the screen as red. Because of this I'm producing the contrasting color green with my eye, just as with the cushion. Thus there isn't any real green. Rather I just see it incidentally because the screen is colored red."

This view of Goethe's is wrong. You can easily persuade yourself that it is wrong, for, if you take a small tube and look through it, so that, after the light has been dimmed, you are looking only at this green band, then you will still see it as green.[1] You're not seeing what surrounds it. Rather you're only seeing the green, which is objectively present at that spot. By doing this, by dimming here and then looking at the green, you can persuade yourself that the green is objective. It stays green and thus can't be a contrast phenomenon; it is an objective phenomenon. We can't make it possible for everybody to see it individually, but "the truth be known when two are shown."[2] I will produce the phenomenon, and you should look through so that you see the green band. It stays green, doesn't it? Just as the other color would stay red if I produced red with green. In this case Goethe incorporated his error into his theory of color, and naturally it has to be corrected.

First, I want you to do nothing more than to keep the facts in mind from the many different kinds of phenomena I have shown you. If we saturate the shadow—gray, in other words, darkness, which otherwise appears merely as a shadow—with color, so to speak, then light and dark interact in a different way than when the shadow isn't saturated with a color. And let's remember that by darkening the light with red here, we cause the objective appearance of green. Now I've pointed out the so-called subjective phenomenon to you. We have a so-called objective phenomenon—the green, which remains on the screen, even if it is not fixed, as long as we create the conditions for it—and then something here, which to a certain extent is subjective and dependent on our eyes alone. Goethe calls the green color that appears to me when I have exposed my eyes to a color for a while the "required" color, the "required" afterimage, which is caused by the reaction itself.

Now here is something to keep firmly in mind. The differentiation between the subjective and the objective, between the color that is temporarily fixed here and the color that is apparently only a color "required" as an afterimage, has no justification on the basis of objective facts. When I am seeing the red here with my eyes, I am dealing simply with all the pieces of physical equipment I have described to you—the vitreous body, the lens, the fluid between the lens and the cornea. I am dealing with a highly differentiated physical apparatus. The relationship of this physical equipment, which mixes light and dark with each other in the most varied ways, to the objectively extant ether is no different than that of the pieces of equipment I have set up here—the screen, the rod, etc. In the one instance, the entire setup, the entire machinery, is just my eye, and I see an objective phenomenon with my eye, only here the phenomenon lasts. If, however, by looking I prepare my eye so that afterward it operates in the so-called subjective required color, then the eye returns in its conditions to a neutral state. However, the process by which I see green is not at all different when I see it in a so-called subjective way with my eyes than when I objectively fix my gaze on it here.

That's why I said you don't live subjectively in such a way that the ether out there makes vibrations whose effect is expressed as color. Rather you float in the ether. You are one with it, and whether you become one with the ether by means of the equipment here, or by means of what takes place in your eyes themselves, is just a different series of events. There is no real, essential difference between the green image that has been produced in space by darkening with red and the green after-image that only appears temporarily. Looked at objectively, there isn't a tangible difference—only, in one instance, the process takes place in space, while in the other instance it takes place in time. That's the only intrinsic difference. If you pursue the essence of these things, it will lead you to understand the opposition of the so-called subjective and objective as it really is, not in the wrong sense in which it is continually understood by the modern natural sciences. In the one instance, we have a setup to produce colors, and our eyes remain neutral; that is, they make themselves neutral vis-à-vis the emergence of color. In other words, the eye becomes one with what is there. In the other instance, the eye itself acts as a physical apparatus. However, whether this physical apparatus is here (outside), or in your frontal sinus is all the same. We are not outside of things and don't just project phenomena into space. We are thoroughly in things with our being and are in things all the more as we ascend from certain physical phenomena to other physical phenomena. No unbiased person who investigates color phenomena can do anything but admit that we are not in them with our ordinary bodily nature, but with our etheric and, therefore, with our astral nature.

If we descend from light to heat, which we also perceive as something that is a condition of our surroundings and which acquires significance when we are exposed to it, we will soon see that there is a significant difference between the way we perceive light and the way we perceive heat. You can localize the perception of light in the physical apparatus of the eye, whose objective significance I have just characterized. In the case of heat, what do you have to say to yourself? If you really ask yourself how you can compare the relationship that you have to light with the relationship you have to heat, you have to answer as follows: My relationship to light is limited by my eyes, so to speak, to a specific place in my body. That's not the way it is in the case of heat. Here I am more or less all sense organ. I am to heat just as the eye is to light. Thus we can't speak in the same localized sense about the perception of heat as we can about the perception of light. However, just by turning our attention to something of this kind, we can arrive at yet another point.

What *are* we actually perceiving when we enter into a relationship with the heat element of our surroundings? Indeed we actually perceive this floating in the element of heat in our surroundings very clearly. Only, *what is floating?* I'd like you to answer that question for yourself: What is it that actually floats there, when you are floating in the heat of your surroundings? Let's take the following experiment. Fill a trough with moderately warm water, with water that feels lukewarm to you when you put both hands in—don't put them in long; just test it.

Then do the following: first put your left hand in water that is as hot as you can stand it, then put your right hand in water that is as cold as possible, and then stick the left and the right quickly in the lukewarm water. You will see that the lukewarm water seems very warm to the right hand and very cold to the left hand. The heated left hand perceives as cold the same thing that the cooled right hand perceives as heat. Before you felt a uniform lukewarm temperature. What part of you is it that floats in the heat element of your surroundings? It is your own heat, which is produced by your own organic process. It isn't something unconscious—your consciousness lives within it. Inside your skin you live in your body heat, and, depending on what this is, you come to terms with the element of heat in your surroundings. Your own body heat floats within it. Your heat organism floats in the environment.

If you think such things through, you will approach the real processes of nature quite differently than by means of what current physics can offer you, which is completely abstracted and removed from all reality.

Now, however, let us descend even further. We have seen that we can say that when we experience our own heat condition, we do so because we float together with it in our heat environment. Thus when we are warmer than our surroundings we experience them as draining us—when the surroundings are cold—and when we are colder, we experience the environment as giving us something. It's quite another thing if we are living in a different element. You see, we can live in that which is the basis of light. We float in the element of light. We have just discussed how we float in the element of heat. But we can also float in the element of air, which we actually have continuously within us. Indeed we are solid bodies only in very small measure. Our bodies are only to a small extent solid. We are actually over 90 percent a column of water, and, particularly in us, water is only an intermediate state between the airy and solid states. We certainly can experience ourselves in the airy element, just as we experience ourselves in the element of heat. In other words, our consciousness effectively descends into the airy element. Just as it enters into the elements of light and heat, our consciousness also enters into the element of air. By entering into the element of air it can in turn come to terms with what happens in the air surrounding us, and this coming to terms is what manifests itself in the phenomenon of sound or tone.

We have to differentiate different levels of our consciousness. With one level, we live in the element of light by taking part in it ourselves. With a different level, we live in the element of heat by taking part in it ourselves. And with yet another level of our consciousness, we live in the element of air by taking part in it ourselves. Because our consciousness is capable of descending into the gaseous, airy element, we live in the airy element of our surroundings and are thus able to perceive acoustical phenomena, to perceive sound phenomena. Just as we ourselves have to take part consciously in light phenomena in order to be able to float in the light phenomena of our surroundings, and we ourselves have to take part consciously in the element of heat in order to be able to float in it—in the same way we also have to take part in the airy element. We must have something airy differentiated within us in order to be able to perceive differentiated airy things, let's say, a pipe, a drum, or a violin.

In this respect our organism presents something extraordinarily interesting. We breathe air out—our breathing process, of course, consists of breathing air out and breathing air in again. When we breathe air out, we push the diaphragm upward. Connected to this, however, the entire system of organs beneath the diaphragm is relieved of strain. Because we raise the diaphragm when breathing out, the cerebrospinal fluid, in which the brain floats, is pushed downward to a degree. However, the cerebrospinal fluid is nothing more than a condensed modification, if I may say so, of the air, for in truth it is the exhaled air that causes this action. When I breathe in again, the cerebrospinal fluid is pushed upward.[3] Thus, by breathing, I live continuously in this upward and downward swinging of the cerebrospinal fluid, which is a clear reproduction of the entire process of breathing. If I live consciously in the fact that my organism takes part in these oscillations of the breathing process, then there is an inner differentiation in my experience of the airy element of consciousness. Through this process, which I have described only roughly, I am placed continuously within a life rhythm consisting, both in its origins and in its course, of the differentiation of air.

What takes place here internally when you breathe—differentiated not so crudely, of course, but in diverse ways—this upward and downward swinging of the rhythmic forces, may itself be described as a complex, continuously rising and falling oscillation organism. We cause this internal organism of oscillation to collide in our ear with sounds coming to us from the outside, for example when a string is struck. And just as you perceive the heat of your own hand when you dip it into lukewarm water, by comparing the difference between the heat of your hand and the heat of the water, you perceive the corresponding tone or sound by comparing your wonderfully constructed inner musical instrument with the phenomenon in the air that manifests itself as tones or sounds. The ear is only the bridge that your inner Apollonian lyre uses to balance itself in its relationship with the differentiated air movements coming to you from the outside. If I portray it in real terms, the actual process of hearing, namely, differentiated sounds or tones, is quite distinct from that abstraction whereby they say that something outside acts to affect my ear and the effect is perceived as an effect on my subjective being—which, in turn, they describe (but with what terminology!)[4] or, actually, don't describe. No matter which idea is used as the basis here, we can't get any further if we want to think things through clearly. We can't come to a conclusion about certain things that are commonly taken up, simply because this kind of physics is far from going into the facts.

In terms of the facts, you are dealing with three stages of the relationships of human beings to the outer world—the light stage, the heat stage, and the sound or acoustic stage. However, there is something very peculiar here. If you examine without bias your relationship with light, that is to say, your floating in the element of light, then you have to say that you can only inhabit the processes of the outside world as an ether organism. By inhabiting the element of heat, you are living with your entire organism in the element of heat of your surroundings. Now direct your attention downward from this aspect of living within these elements into the inhabitation of the sound and tone element. In this case, by becoming yourself an organism of the air, you actually inhabit differentiated forms of the outer air. In other words, you no longer inhabit the ether, but actually live in the external physical substance—you inhabit the air in this case. For this reason, life in the element of heat is a significant dividing line. To a certain extent, the element of heat, living within it, means a middle level for your consciousness.

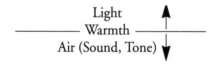

You can perceive this level very clearly in the fact that for all intents and purposes you can hardly distinguish outer and inner heat in terms of pure sensation. However, life in the element of light lies above this level. You ascend into a higher etheric sphere, so to speak, in order to inhabit it with your consciousness. And you penetrate below that middle level, where you balance yourself out with the outer world in relatively simple fashion, by coming to terms with the air in the perception of tones or sounds.

If you put everything I have just shown you together with what I have said about anatomy and physiology, then you can't help but regard the eye as a piece of physical equipment. The farther you move toward the outside, the more physical you find the eye to be. The farther you move inside, the more it is permeated by vitality. Thus we have within us a localized organ to raise us above the middle level. On this middle level we live on equal terms with the environment when we approach it with our heat and perceive the difference somewhere. In this instance we have no specialized organ such as the eye—here, in a way, we ourselves turn completely into a sense organ. Now we dive down under this level. In the element where we turn into beings of the air, where we come to terms with the differentiated outer air,[5] this conflict is localized once again. Something is localized there, between the outer air and the process that is taking place within us, in this Apollonian lyre, this rhythmic play of our organism, which is simply reproduced in the rhythmization of the cerebrospinal fluid. What takes place there is connected by a bridge. Once again there is a localization of this kind, but now it is below this middle level, just as in the case of the eye we have such a localization above this level.

Our psychology, you see, is actually in even a worse state than our physiology and physics, and we can't really blame the physicists very much for expressing themselves so unrealistically about what is in the outer world, because they are not supported at all by the psychologists. The psychologists have been conditioned by the churches, which have staked a claim to all knowledge about the soul and spirit. Therefore, this conditioning, which the psychologists have accepted, has led them to regard the human being as only the outer apparatus and to see soul and spirit only in the sound of words, in phrases. Our psychology is actually only a collection of words, for there's nothing there about what people should understand by "soul" and "spirit." And that's why it appears to the physicists that it is an inner, subjective experience when light at work out there affects the eye and the eye counteracts it or receives the impression, as the case may be. A whole tangle of ambiguities begins right there, and the physicists repeat this in quite the same way for the other sense organs. If you read through books on psychology these days, you'll find it to be a theory of sense perception. They speak of sense, of sense in general, as if there were such a thing. Just try studying the eye. It's something quite different from the ear. I have characterized that for you—how they lie above and below this middle level. The eye and the ear are organs with an inner formation of quite different kinds, and that is what we should take into consideration in a significant way.

Let's stop here for the moment. Think that over, and tomorrow we'll talk about acoustics, the theory of sound, from this point of view so that starting from there you will be able to master, in turn, the other areas of physics.

Today I would like to show you just one more thing. It is something that in a certain respect we could call the pièce de résistance of modern physics; it is indeed in a certain respect a pièce de résistance. If you simply stroke a surface with your finger, thus exerting pressure by your own effort, the surface becomes warm. By exerting pressure you have produced heat. By creating objective, mechanical processes, processes that are definitely mechanical, we can once again produce heat. So, as another basis for what we want to look at tomorrow, we have improvised this apparatus. If you checked the temperature in this apparatus, you would get a reading of 16° Celsius, more or less. Now we have water in this receptacle, and in this body of water there is a flywheel, a drum that we cause to turn rapidly, so that it does mechanical work, stirring up the water thoroughly, scooping the water up. Then, after a bit, we'll look at the temperature. You will see then that it has risen. Thus the temperature of the water has increased purely by mechanical work. In other words, heat is produced by mechanical work. This was worked up then, initially in the form of calculations, after Julius Robert Mayer,[6] in particular, pointed it out. Julius Robert Mayer himself formulated it as the so-called mechanical equivalent of heat. If it had been developed further according to his ideas, we would have said nothing more than that a certain number is the expression for the measurable heat divided by mechanical work, and vice versa. However, this was analyzed in a supernatural, metaphysical way when it was stated thus: if there is a constant relationship between the work that has been done and the heat, then the latter is simply transformed work. Transformed!—whereas we are dealing for the moment with nothing but the numerical expression for the connection between mechanical work and heat.

Eighth Lecture

STUTTGART, DECEMBER 31, 1919

CONVENTIONAL DESCRIPTIONS of physics have described tone and sound in the current manner only since the fifteenth century, more or less. It's just such examples that corroborate what I've often expressed as an insight of spiritual science— that people's entire way of thinking and imagining was different before the turn of this age than it was after. Moreover, this way of speaking about sound phenomena only gradually took on the academic form in which we speak today. The first thing people became aware of was the speed with which sound is propagated. It is relatively easy to get an idea of the speed of sound propagation, at least with a certain approximation. If you shoot off a cannon at some distance, you see the flash of the light phenomenon far off and afterward hear the bang, just as you hear thunder later than you see lightning. If you ignore the speed of light, the time that passes between the perception of the light impression and the perception of the sound can be regarded as the time it took the sound to travel the given distance. Then you can calculate how quickly sound travels in air, say in one second, and arrive at something like the speed of the propagation of sound.

That was one of the first aspects people became aware of in this area. What we call resonance also drew their attention, especially that of Leonardo da Vinci.[1] Today you know this as sympathetic vibrations: if a string is struck in a space where there is another string or another object with the same tuning, the other string or object will vibrate sympathetically. The Jesuits in particular studied such things. For example, in the seventeenth century the Jesuit Mersenne[2] accomplished a great deal for the theory of sound or tone in connection with studies on pitch. You can distinguish three elements of a sound: first, the sound has a certain loudness; second, it has a certain pitch; and in addition, it has a certain quality. Of the three, the most important, the most essential is the pitch. Now, it's a matter of determining what the source of pitch is from the point of view that gradually developed, particularly in the study of sound. As I've already pointed out, when we perceive a sound, the cause or, let's say, the concurrent phenomenon is something that is vibrating. It's very easy to establish the vibrating condition of the air or other bodies by the usual experiments. It's not necessary to discuss these experiments in detail—all you have to do is imagine yourself back in school. You strike something like a tuning fork and then follow the line with a pencil attached to it.[3] In the image it reproduces in soot you will see that the tuning fork is moving regularly. It goes without saying that this regular motion is conducted to the air. Thus, it can be said that whenever we hear a sounding body, the air between it and us is in motion. We put the air directly into motion, of course, in the devices that we call pipes.

People gradually arrived at an understanding of what kind of motion is involved. We are dealing with so-called longitudinal vibrations. The fact that we are dealing with longitudinal vibrations can also be proven. You generate a tone in a metal pipe and connect the pipe with a tube filled with air so that it conducts the motions of the metal pipe. If to the air-filled tube you add dust that is easily moved, you will be able to ascertain from the motion of the particles of dust that the sound continues as follows: First, there is compression of the air. This compression of the air is pushed back, in turn, when the body vibrates back, thereby producing an expansion of the air. At the moment when the metal moves forward, the original compression moves forward. Compression and expansion alternate in this fashion. Thus it is experimentally possible to prove that compression and expansion are involved. It really isn't necessary for us to carry out such experiments, since these things are obvious, if I may say so. I don't want to present you with anything that can be gotten out of books.

At the beginning of the modern era the Jesuits, with the help of their social connections, accomplished an extraordinary amount, particularly for such branches of physics. However, they always endeavored not to penetrate natural processes spiritually in any way or to observe the spiritual element in natural processes, but to reserve spirituality for religious life. The Jesuit side always regarded it as dangerous to apply to natural phenomena a "spirit-suited" method of looking at things—to use the expression we are accustomed to hearing from Goethe.[4] The Jesuits wanted to look at nature purely materialistically, in no way approaching nature with the spirit, and in many respects they were the first keepers of the materialistic views that are particularly dominant today. We don't think about the fact, although we know it historically, that the way of thinking employed in today's physics is basically a product of this Catholic tendency.

Now we are chiefly concerned with why we hear sounds with different pitches. In what ways do the differences in external vibration phenomena that appear in sound relate to the differences in pitch? We can show these things by experiments like the one that we are going to demonstrate. We will cause this disk with different rows of holes to move rapidly, and Mr. Stockmeyer[5] will then be so kind as to direct a stream of air at this moving disk. [This is done.]

You can easily distinguish how the pitch differed. What caused this difference? It was caused by the fact that we have a smaller number of holes on the inner part of the disk—only forty holes. When Mr. Stockmeyer directed the stream of air at the disk, the stream went through when it came to a hole but wasn't able to go through between the holes, and so forth. Because of the movement of the disk, the next hole arrived in place of the previous one, and in this way as many gusts of air were caused as holes arrived at the spot where the stream of air was passing. Because of this we have forty gusts of air on the inside here and on the outside circle we have eighty gusts of air. The gusts of air cause waves, vibrations. Therefore in the same time period—because these eighty holes complete a turn in the same amount of time as the forty inner holes—we have eighty gusts of air one time, eighty vibrations of the air, and forty gusts of air, forty vibrations of the air, the other time. The sound that results when we have eighty vibrations is twice as high as the one that results when we have forty vibrations. By these and similar experiments we can prove that pitch correlates to the number of vibrations that take place in the medium in which the sound is propagated.

If you take what a vibration consists of—in other words, one compression and one expansion—we can call that the wavelength. Now, if in one second n waves of the length l are generated, then the whole wave movement progresses at the rate of n times l. That is, the path that the whole wave movement travels in one second—I'll call it v—is n times l. Now I'll ask you to remember what I referred to in my previous reflections. I told you that we have to carefully distinguish everything that is kinematic from those things that have not simply been invented by the inner life of the imagination but are external realities. I also told you that external realities can never exist solely in terms of number, space, and movement; speeds are always external realities. Naturally it is no different when we speak of sound. Outward experience doesn't consist of either l or n, for l is merely spatial and n is merely a number. It is in speed that reality lies, and when I divide into two abstractions the speed contained within the being that I call sound, then naturally I don't get any actual realities. Rather I get what has been abstracted, separated, and compartmentalized. Wavelengths, spatial measurements, and the number n are all such compartmentalized abstractions. If I want to look at the reality of sound, at what is real externally, I have to look at the inner ability of sound to have speed. That is what leads to a qualitative examination of sound, while the view we are accustomed to in physics today is a quantitative examination of sound. It is particularly striking that in acoustics what can be noted in terms of space, time, movement, and number is nearly always substituted for the qualitative, which expresses itself solely in terms of a certain capacity for speed.

These days we no longer notice at all how even in acoustics we have basically wandered off into dangerous materialistic territory. We can say that it's so obvious that sound as such just doesn't exist outside of us. Outside of us there are simply vibrations. How could anything be clearer than this? When a stream of air is produced, which creates compressions and expansions, and my ear hears them, then that unknown something within me (which physicists don't have to go into, of course, because that's not physics) converts the air vibrations into purely subjective experiences, converting the vibrations of bodies into the qualitative nature of sound. And you will find expressed in the most diverse ways the idea that outside of us vibrations exist, but that inside of us are the effects of these vibrations, which, however, are purely subjective. This idea has gradually become second nature to us, with results like those you can find cited from Robert Hamerling's works in my *Riddles of Philosophy*.[6] Hamerling, in taking up the theories of physics, says right from the outset that what we experience as an explosion is nothing but a disturbance of the air outside of us; and whoever can't believe that what we actually experience as a sensation is only within us, and that externally there is just vibrating air or vibrating ether, shouldn't read any further in a book such as those written by Robert Hamerling. Hamerling even states that someone who believes that a picture of a horse really corresponds to an outer reality doesn't understand anything and should shut his book.

But my dear friends, we have to follow such things to their logical conclusion. Imagine if I were to treat you who are sitting here in accordance with this physical way of thinking ("way of thinking," I say, not "method") that physicists are accustomed to applying to sound and light phenomena. Then all of you sitting before me would, of course, be in front of me only by means of my impressions, which are completely subjective, like sensations of light and sound. Indeed outside of me none of you would exist as I see you. Instead it would only be the vibrations of the air between you and me that lead me to the vibrations that, in turn, are in you. This actually would bring me to the point where your entire inner soul life, which of course as far as you are concerned is undeniably in you, wouldn't really exist. Instead, for me the inner soul life of all of you who are sitting here would merely be the effect on my own psyche. Otherwise, there would merely be some heaps of vibrations sitting on the benches here. It's the same kind of thinking if you deny light and sound the inwardness that you apparently experience subjectively. It's exactly the same as when I have you here before me and regard what I have in front of me only as something subjective inside of me, denying that you experience this same inwardness.

What I am saying now is apparently so obvious and banal that physicists and physiologists naturally can't imagine committing such errors. But they do it anyway. This whole distinction between the subjective impression—what is supposedly subjective—and the objective process is nothing else. Naturally they could go about things honestly and say, "As a physicist, I don't want to investigate sound at all. I don't want to go into the qualitative aspects. I want to leave that alone and investigate only the external/spatial—let's not say 'objective'—processes, which nevertheless continue on into me. I want to separate them as abstractions from the totality, and I am not getting involved with the qualitative." Then they would indeed be honest, but they shouldn't then assert that this is something objective and that is something subjective, or even that one is the effect of the other, for what you experience in your soul is not—if I share in the experience—the effect of your brain vibrations on me. To meet the demands of modern times and modern science, it is critical to understand these things.

In other words, with such things we can't avoid delving into deeper connections. We can easily say, for example, that the purely vibratory nature of sound follows from the fact that, if I pluck a string, another string in the same space and tuned to the same pitch will vibrate sympathetically. That is simply based on the fact that the intervening medium propagates the accompanying vibrations. However, we can't understand what we observe here if we don't grasp it as part of a much more general phenomenon. The following is the more general phenomenon, which indeed has also been observed.

Let's assume you have a pendulum clock that you have gotten going, and in the same room there is another pendulum clock—it has to be constructed, mind you, in a certain fashion—which you don't start going. You will discover that occasionally, when the conditions are favorable, the second pendulum clock will start running by itself. This is something we can call the sympathy of phenomena, and it can be investigated in broad areas. The last of this type of phenomena that still has something to do with the outer world is one that could be investigated much more than it usually is because it actually occurs often. You can experience it countless times: you are sitting at a table with a person who says something you have just thought. You thought it, but you haven't said it, and the other person expresses it. This is the sympathetic coincidence of events, of connected events that are somehow attuned to each other, which in this case manifests itself in a highly spiritual area. And we will have to discern a continuous series of facts between the simple sympathetic vibration of a string—which, in accordance with unsophisticated notions, we regard materialistically as merely another occurrence fitting into external material events—and those parallel phenomena that manifest themselves more spiritually, such as the coexperience of thoughts.

But we won't be able to get clear insights into these things unless we are willing to deal with how the human being also fits into what we call physical nature. You'll remember that we showed the human eye a few days ago and analyzed it a bit. Today we will show the human ear. As you know, of course, in the back of the human eye is the vitreous body, which we can say still has vitality, and then there is the fluid between the lens and the cornea. As we go from the outside to the inside, the eye becomes more and more alive. Outside, it is more of a physical nature. We can also describe the ear in the same way that we can describe the eye, saying in a superficial way that just as light makes an impression on the eye by affecting the eye (or however you want to say it) so that the nerve receives the stimulus, the sound vibrations act upon the ear, enter the auditory canal, and beat on the eardrum, which closes off the auditory canal. Placed on the back of the eardrum are the ossicles, the hammer, anvil, and stirrup, named after their shapes. Thus in physical terms, what originates and expresses itself externally in the air in the form of compression and expansion waves is transmitted by this system of ossicles to what is located here, in the inner ear. Here in the inner ear is first what we call the cochlea, which is filled with a fluid and in which the auditory nerve ends. Placed on it in front are the three so-called semicircular canals, whose surfaces are perpendicular to each other in the three dimensions of space. So you can imagine it as follows. Sound in the form of airwaves penetrates here. It progresses through the ossicles and reaches the fluid. There it reaches the nerves and acts upon the sensing brain. So there we have the eye as a sense organ and the ear as another sense organ. In this way we can look at these two things neatly juxtaposed, and as a further abstraction we can determine a common theory of the physiology of sense perception.

However, the matter won't seem so simple if you take what I said about the combined effect of the whole rhythm of the rising and falling cerebrospinal fluid and put it together with what happens externally in the air. For, as you will recall me saying, we shouldn't simply assume that something that appears superficially to be self-contained is a complete reality. It isn't necessarily a complete reality. The rose that I break off the rosebush is not a reality, for it can't exist by itself. It can achieve an existence only by virtue of its connection to the rosebush. In truth it's an abstraction if I think about it merely as a rose. I have to go on to the totality, at least to the whole rosebush. Thus, in hearing, the ear is not at all a reality—the ear that is usually described. What is propagated from outside to the inside through the ear has to interact with the process of inner rhythm manifested in the rising and falling of the cerebrospinal fluid. Thus we extend what is happening in the ear to what is happening within these rhythmic movements of the cerebrospinal fluid.

Even then we're not finished. What happens rhythmically in the human being, including the brain in its domain, is in turn the foundation of something that manifests itself in a completely different aspect of our organism, in the act of speaking, by means of the larynx and the neighboring organs. Of course your active speech is, in terms of its tools, integrated into the breathing process, which is also the foundation of this rhythmic process of the rising and falling cerebrospinal fluid. On the one hand, you can just as well include your speech process as part of everything that takes place rhythmically in you when you breathe, and, on the other hand, you can include hearing. Then you have a whole, which manifests itself in the act of hearing more in terms of the intellect and in the act of speech, more in terms of the will. You only have a whole if you combine the will element, which is pulsing through the larynx, with the more intellectual and sensual element, which passes through the ear. They belong together—we have to see this clearly as a simple fact. If we take either the ear or the larynx out of context, it is only an abstraction. You will never arrive at a whole if you separate things that belong together from each other. The physiological physicists or physical physiologists who study the ear and the larynx in isolation are acting, in terms of their research procedures, in exactly the same way that you would be, if, in order to heal a person, you dissected him or her, instead of studying things in their organic interaction.

Once we have a proper grasp of what we are concerned with here, we come to the following. Observe what is left of the eye when I have taken away the vitreous body and all or part of the retina [cf. Figure 3f]. If I could remove all of this, then the ciliary muscle, the lens, and the outer fluid would remain. And what kind of organ would that be? If I work realistically, that would be an organ I would never compare to the ear. Instead, I would have to compare it with the larynx. That is not a metamorphosis of the ear; it's really a metamorphosis of the larynx. To give you only the most general idea, just as the laryngeal muscles grasp the vocal cords, making a wider or narrower slit, the ciliary muscles do the same here. They grasp the lens, which is inwardly mobile. I've separated out the elements that are larynxlike for the etheric, in the way that our larynx is larynxlike for the air. And if I put first the retina and then the vitreous body back in—and with certain animals I would have to put in certain organs such as the pecten, which is present only etherically in the human being, or the falciform process[7]—in certain lower animals these extend inward as vascular organs. If I include all of that, then I can only compare it with the ear. I would compare such things as the spreading parts of the pecten with the spreading parts of the labyrinth in the ear, and so forth. Thus at one level in the human organism I have the eye, which internally is a metamorphosed ear that is enclosed externally by a metamorphosed larynx. On the other hand, if we take the larynx and ear together as a whole, then at another level we have a metamorphosed eye.

I've indicated something that leads in a very important direction. We can't know anything at all about these things if we compare them with each other in a completely false way simply by placing eye and ear next to each other. On the contrary, when comparing the eye with the ear I should look only at what lies behind the lens in the eye, which has more vitality, while in the case of the human larynx I have to compare what pushes out in front here and is more of a muscular nature. That's what makes the science of metamorphoses difficult. You can't look for the metamorphoses in a simplistic way; you have to investigate the inner dynamic, the reality, the actuality.

However, if that is the case, we are constrained from comparing sound and light so easily as parallel phenomena. If we start out from the false premise that the eye and the ear are equally sense organs, then we will have a completely false view of what follows from this relationship. When I see, it is something quite different from when I hear. When I see, the same thing happens in the eye as when I hear and speak at the same time. On a higher level, an activity that I can only compare to speaking accompanies the actual receptive activity of the eye. We can really achieve something in this area only if we try to grasp the realities. For when we become aware that here in the eye two different kinds of things are united, which otherwise, in the case of hearing, of sound, are shifted to apparently completely different bodily organs, then we start to realize that in seeing, in the eye, something takes place that is like a kind of communication with oneself.

The eye always acts the way that you act when you hear something and repeat it first, in order to understand it. The nature of the eye's activity is really as if you listened, but didn't grasp it correctly yet. If the other person says, "He is writing," you're still unclear, so you repeat, "He is writing." Only then is the whole matter complete. That's the way it is for the eye with light phenomena. They enter our consciousness because of the curious fact that our eye has a vital part, but this becomes the full experience of sight only when we reproduce it in the part of the eye, located in the front, that corresponds to the larynx. When we see, we are speaking with ourselves etherically. The eye is talking to itself. Thus we can't compare something that is the result of talking with ourselves—in other words something that already comprises the typical activity of the human being—with hearing alone, which is only one element, one part.

If you thoroughly think this idea over for yourself, you can gain an extraordinary amount from it, because you can see how far the materialistic view of the world has wandered off into absolute unreality by comparing things that are not directly comparable at all, such as the ear and the eye. Because of the very superficiality of this way of viewing things, which does not look at real wholes, we stray from a spiritual view of nature.

Consider to what degree the conclusion of Goethe's *Theory of Color*, in the sensory-moral section, derives the spiritual from the physical. And you can never do this if your basis is the current physical theory of color.

Now, however, we begin to have doubts about sound, whether it is as self-evident as they say that externally sound is only a matter of vibrations. But you have to ask yourself this question—and I ask you to decide for yourself whether this question isn't already answered to a certain extent by being asked in the right way: Could the following be the case? Let's say that you have a globe here filled with air, and that you also have a hole in the globe that can be opened by a valve. Nothing will happen as long as the air on the inside has the same density as the air on the outside—even if you open the hole. But if there is a vacuum in the globe, then something will indeed happen: the outer air will whistle in and fill up the empty space. Would you by any chance say in this case that the air that is inside afterward has simply been created by what happened inside? No, you would say that the air penetrated from outside; however, based purely on observation, the empty space sucked the outer air in.

When we start the disk spinning and shoot air through here, we simply create the conditions for something to happen, which we have to call a sucking effect. What emerges afterward in the form of sound when I set the siren in motion and cause the air to vibrate—that is beyond the space; it is not yet inside the space. The conditions aren't there for it to come into the space until I produce them, just as the conditions aren't there for the outer air to rush in here until I produce them. I can only compare the external air vibrations to the vacuum, and I can only compare the audible sound to the outside air that penetrates the vacuum because the conditions have been created for it. But intrinsically, in their essence, the vibrations of the air have nothing to do with the sound except that where these air vibrations are a vacuum effect takes place, which pulls the sound in. It goes without saying that the nature of the sound that is pulled in will be modified depending on the type of air vibrations, but it would also be modified here in the vacuum, if I were to make passages here and the air expanded along certain paths. Then these passages would be a copy of the lines along which the air expands. Likewise sound phenomena are copied in the vibration phenomena that take place externally.

So you see that it isn't so easy to imagine how by means of a few mathematical ideas about vibration phenomena we can indicate the basis of a real physics of sound. It makes more demands on the qualitative in human thinking. But unless we fulfill these demands adequately, we will create only that construct of a physical conception of the world—the physical worldview that we worship today—that relates to reality the way a papier-mâché person relates to a living human being. Think that over, and we'll continue next Friday.

Ninth Lecture

I'M TERRIBLY SORRY that these discussions have to be so very improvised and aphoristic, but there's just no other way than to give you a number of points and then continue the subject when I am here once again sometime soon. In the course of time you will be able to get something complete out of these discussions. However, I want to give you a couple of points that I will develop for you in closing tomorrow, which, in turn, will make it possible to throw some light on the pedagogical application of scientific insights. Today I want to direct your attention to the development of electrical phenomena, taking up some things that are actually already familiar to you from your school days, because tomorrow, starting from there, I want to give you an overview of the entire area of physics.

You already know the elementary facts about the theory of electricity. You know about the existence of what we call static electricity—that we can cause a glass or resin rod to develop a force by rubbing it with some kind of so-called friction cloth, so that the rod becomes electric, so to speak, meaning that it attracts small bodies, such as little bits of paper. You also know that observation of the phenomena has gradually shown that the force exerted by the glass rod on the one hand and that exerted by the rod made of resin or sealing wax on the other exhibit different properties. When the glass rod attracts the paper cuttings, they are saturated electrically, as they say, in a particular way that is opposed to the way they are saturated electrically by the resin-rod electricity. Therefore, with a nod to the qualitative, we differentiate between glass electricity and resin electricity, or, expressed more generally, positive electricity and negative electricity. Glass electricity is positive; resin electricity is negative.

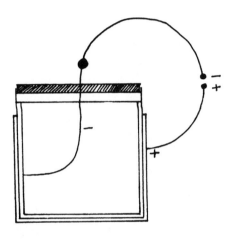

Figure 9a

Now the curious thing is that positive electricity somehow always attracts negative electricity. You can observe this phenomenon in the so-called Leyden jar. This is a vessel coated on the outside with an electrical conductor, which is isolated here, and coated on the inside with another conductor, which is attached to a metal rod with a metal knob [Figure 9a]. If we electrify a metal rod and conduct the electricity to the outer coating—which you can do—then the outer coating will become, for example, electrically positive, producing the phenomena of positive electricity, with the inner coating becoming electrically negative. Then, as you know, if we connect the coating that is imbued with positive electricity with the coating that is imbued with negative electricity, by creating the situation that the one electricity can continue to here and oppose the other one, we can produce a connection between the positive and negative electrical forces. They oppose each other with a certain tension and demand its resolution. A spark jumps from one coating to the other. Thus we see that opposing electrical forces have a certain tension that strives toward resolution. I'm sure this experiment was often done for you.

Here you see the Leyden jar. But we also need a discharging fork. Now I'm going to charge it here. It's still too weak. The little plates are repelling each other a little. If we were to charge it sufficiently, the positive electricity would evoke the negative electricity, and, if we had them opposite each other, with a discharging fork we would cause a spark to jump over. As you also know, this method of electrification is called static or frictional electricity, because we are dealing with a force of some sort that has been produced by means of friction. For the moment, that's how I would put it.

I need only remind you that we found our way to static electricity and also discovered what we call galvanic electricity only around the turn of the nineteenth century. This opened up territory that has proven to be extraordinarily fertile for the materialistic development of modern physics. You need to recall the principle involved. Galvani[1] observed that a frog's leg connected to metal plates began to twitch, thereby discovering something extraordinarily significant—if I may say so, actually two things at the same time. However, these needed to be differentiated from each other, but they have not been properly differentiated to this day, much to the harm of scientific study. Galvani had discovered what Volta[2] was able to describe a little later as "the true contact electricity": the fact that if two metals touch each other in such a way that their contact is mediated by liquids of the right kind, then a reaction takes place that assumes the form of an electrical current from one metal to the other. The result is an electrical current that apparently acts purely in the inorganic realm. However, when we look at what Galvani actually discovered, we also have something that we can describe as "physiological electricity," a tension of forces that always prevails between muscle and nerve and that can be awakened when electrical currents are passed through them. Therefore what Galvani saw at that time was in fact two different things: the phenomenon that we can reproduce easily in the inorganic realm by causing metals to bring forth electrical currents with the mediation of liquids; and the phenomenon that is present in every organism, but manifests itself particularly in certain electric fish and other animals, as a state of tension between muscle and nerve, which superficially looks like flowing electricity and its effects in the way it discharges. Thereupon everything had been discovered that afterward led to tremendous advances in materialistic scientific knowledge on the one hand and resulted in such enormous, epoch-making foundations for technology on the other.

The point is that the nineteenth century was for the most part consumed by the idea that we had to discover some abstract uniformity underlying all natural forces, as they are called. The things brought to light in the 1840s by Julius Robert Mayer, the well-known and brilliant Heilbronn physician I have already mentioned, had also been interpreted in this way, of course. We've already demonstrated what he brought to light: by turning a flywheel we generated mechanical force, causing the water to become mechanically active internally. Because of this it became warmer. We were able to prove that the water warmed up, and it's possible to say that the development of heat is an effect of the mechanical effort, of the mechanical work that was done. These things were interpreted in such a way that they were applied to a great variety of natural phenomena, which of course within limits could easily be accomplished. It was possible to cause chemical forces to develop and to see how heat resulted from the development of chemical forces. Conversely, it was possible to use heat to produce mechanical work, as happens in the steam engine in the fullest sense.

The so-called conversion of natural forces attracted special attention, brought about by further developments of what began with Mayer: the fact that we can calculate how much heat is necessary to accomplish certain measurable work and, conversely, how much mechanical work is necessary to produce a certain measurable quantity of heat. Although there was at first no reason for it, it was thought that the work done when a paddlewheel turned in the water was simply converted, that this mechanical work changed into heat. It was assumed that when heat is used in the steam engine, the heat is converted into the resulting mechanical work. This was the direction of thought taken by the pondering about physics in the nineteenth century, which therefore strove to find the relationships between the various so-called natural forces, relationships that were supposed to show that in reality some abstract equivalence actually resided in all these different natural forces.

This striving reached a pinnacle of sorts toward the end of the nineteenth century when, with a certain ingenuity, the physicist Hertz[3] discovered the so-called electrical waves—once again it was waves. This provided some justification for thinking the phenomenon of electricity was related to the phenomenon of light, which of course was conceived of as a wavelike motion of the ether. The fact that what we were dealing with in electricity, especially in the form of electrical current, wasn't to be understood so easily by primitive, mechanical, basic concepts, but actually made it necessary to expand the horizons of physics to include the qualitative, could have been shown by the presence of what are called induction currents. To give you only a rough idea of this, an electrical current moving in a wire induces a current in a wire nearby simply because the two are in the same vicinity. Thus we could say more or less that effects of electricity take place across space.

Hertz arrived at the interesting insight that the transmission of electrical forces is indeed related to all the phenomena that are propagated in the form of waves or can be thought of in this way. He found that if an electrical spark is produced in the same way it is produced here, that is, if a voltage is generated, then the following result will be achieved. Let's assume that we had a spark jumping across here. Then we could place two such things—let's call them little inductors—opposite each other; they would just have to be placed facing each other in a certain position. At an appropriate distance a spark could jump across here too, which would resemble no other phenomenon so much as one where, let's say, a source of light is here, and a mirror here that reflects the beam of light, which is caught by another mirror here, with the image then appearing here [Figure 9b]. We can speak of the spreading out of light and of an effect that takes place at a distance.

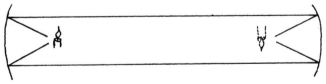

Figure 9b

So Hertz could also speak of electricity spreading out, with its effect perceptible at an appropriate distance. And in his opinion and that of others, he had brought about something that was evidence that with electricity something actually spreads out in a way that corresponds to a wavelike motion—however we might conceive of wavelike motions spreading out. Just as light spreads out through space and acts at a distance when it strikes other bodies and is able to manifest itself, electrical waves also spread out and manifest themselves at a distance. That is the basis for so-called wireless telegraphy, as you know. Thus we are dealing with a certain fulfillment of an idea that was near and dear to the physicists of the nineteenth century: wave movements, which were imagined in the case of sound and in the case of light, and had begun to be imagined in the case of heat transmission because heat phenomena reveal similar characteristics, could also be imagined in the case of electricity; you just had to imagine very long waves. To a certain extent this delivered irrefutable proof that the way of thinking of physics in the nineteenth century was fully justified.

Nevertheless there is something in Hertz's experiments that indicates the old way has actually come to a close with them. You see, everything that takes place in a certain domain can only be judged appropriately within that same given domain. When we have experienced revolutions recently, they have seemed to us like tremendous disruptions of social life, simply because we are looking at their particular domains. If we look at what happened in the realm of physics in the 1890s and in the first fifteen years of this century, we have to say that a revolution has taken place there that is actually much greater in its domain than the external revolution in its domain. Physics is basically in the midst of a complete dissolution of the old physical concepts, although the physicists resist admitting it. While Hertz's discoveries are still the sunset of the old way, because they served to confirm the old wave theory, what came later—and which was already present in Hertz's time, in a preparatory stage, so to speak—has had revolutionary significance for physics.

It consisted of the following. Electrical current, which can be produced and passed on, is conducted through glass tubes from which the air has been pumped out to a certain degree. In other words, the electrical current is conducted through air that is extremely rarefied. You see here that the electrical potential is achieved simply by extending the ends where the electricity can discharge as far apart as the length of this tube so that the part (which we can call one end) through which the positive electricity is discharged, the positive pole, is on one side, and the negative pole is on the other end. The electricity is discharged between these two ends, and the colored line you see here is the path taken by the electricity. Thus we can say that what otherwise passes through wires assumes the form you see here when it is propagated through rarefied air. It is even stronger with more highly rarefied air. Here you see that a kind of movement takes place from one side and the other when there's a significant change in the phenomenon. Thus it is possible for us to manipulate along part of its path what flows through the wire in the form of electricity in such a way that by interacting with something else, it reveals something of its inner essence. It reveals itself because it can't hide in the wire. Take a look at the light in the glass! That is fluorescent light.

I'm sorry that I can't describe these things in more detail, but I wouldn't get to what I want to accomplish if I didn't speak in such a sketchy way.

You can see what is passing through there in a very dispersed state in the highly rarefied air in the tube. Now the phenomena that are revealed in this way in rarefied air or gas tubes need only to be studied. People of all sorts have taken part in these studies, among them Crookes.[4] It's a matter of observing how the phenomena in the tube actually behave and of conducting experiments with the phenomena that result in the tube. Certain experiments also carried out by Crookes, for example, showed that what manifests itself there where we have laid it bare as the inner character of electricity, if you like, can't have anything to do with something that is propagated by means of wave movements of the ether in the way that people wanted to imagine that light is propagated. For what is shooting through the tube here has unusual characteristics that are strongly reminiscent of the characteristics of something that is simply material. If you have a magnet or an electromagnet (I have to appeal to what you already know; we can't discuss everything today), you can attract material objects with it. This body of light, this modified electricity that is passing through there, can also be attracted by a magnet. It behaves toward a magnet in much the same way as matter behaves toward a magnet. The magnetic field modifies what goes shooting through there.

These experiments and others like them led Crookes and other people to think that what's in there cannot be called a continuous wave movement in the traditional sense; instead there are particles of matter in there that shoot through space and are attracted by magnetism like particles of matter. For that reason, Crookes called what was shooting across here "radiant matter," and he thought that because of the rarefaction, the matter that is inside the tube gradually reached a state in which it is no longer only a gas. Instead it becomes something that goes beyond the gaseous state—radiant matter, that is—matter whose individual particles ray out through space like finely dispersed dust, so to speak, and, because of the electrical charge, have the characteristic of shooting through space. Accordingly, these particles themselves are attracted by electromagnetic force. The fact that they are attracted is precisely the proof that we are dealing with the last remnants of real matter, not merely with a motion like the traditional conception of etheric movement. These experiments could be carried out particularly with the radiation that resulted from the negative pole, the so-called cathode. The studies of these radiation phenomena of the cathode, called cathode rays, struck the first blows, so to speak, against the old physical conception. What took place in Hittorf's tubes[5] proved that we are actually dealing with a material thing that is shooting through space, albeit in a very finely dispersed state. Of course, this didn't decide the question of what made up this thing called "matter," but in any case it indicated that it had to be identified as something material.

Thus it was evident to Crookes that he was dealing with some material thing scattering through space. This view undermined the old wave theory. On the other hand, there were other experiments that, in turn, didn't justify Crookes's point of view. In 1893, for example, Lenard[6] succeeded in diverting these so-called rays emanating from this pole—it is indeed possible to divert them—and he was able to conduct them toward the outside, interposing an aluminum wall and conducting the rays through it. The question then arose whether it could be so easy for particles of matter simply to pass through a material wall. This brought up the question, Are these really particles of matter flying through space, or is it something else that's flying through space here after all? Now, you see, that led gradually to the insight that neither the old concept of vibrations nor the old concept of matter was getting us any further. With Hittorf's tubes we were able to follow electricity along its secret paths, so to speak. There had been hope of finding the characteristics of waves, but we weren't able to find them. Then we consoled ourselves with the idea that it had to be matter shooting through space, but even that didn't go right. Finally, as the result of a great variety of experiments, it was said that what is present isn't vibrations or any kind of dispersed matter; rather, what is present is moving, flowing electricity. The electricity itself is flowing, but when it flows, it shows certain characteristics like those of matter in its behavior toward other things, a magnet for example. Naturally, if you make a bullet shoot through space and have it pass by a magnet, it will be diverted from its path. Electricity does the same thing, speaking for the fact that it is something material. But since it can pass easily through an aluminum plate, on the other hand, it reveals itself as something that is not matter. Matter, of course, makes a hole, for example, when it passes through other matter. Thus we ended up calling it "flowing electricity."

This flowing electricity revealed the most peculiar things, and, if I may say so, the direction the studies took made it possible to make the most peculiar discoveries. For example, we were gradually able to investigate how currents that meet the cathode rays are emitted from the other pole as well. This end is called the anode, and the rays it emits are called anode rays. So it was believed that there were two rays that met each other in such a tube.

In the 1890s, something particularly interesting resulted when Röntgen[7] directed cathode rays—captured them, we could say—on a screen he placed in their path. When we capture the cathode rays with a screen, we get a modification of these rays. They continue in a modified form, and we get rays that act to charge certain bodies with electricity and that also interact with certain magnetic and electrical forces. We get what we have gotten accustomed to calling Röntgen rays or X rays. Yet more discoveries followed. You know that X rays have the characteristic of being able to pass through bodies without causing perceptible disturbances and of going through flesh and through bones in different ways, so that they have become very significant for physiology and anatomy.

Then a phenomenon arose that made further thought necessary. When these cathode rays or their modifications strike glass bodies or other bodies—the substance that because of a certain theoretical chemical background is called barium platinocyanide, for example—a certain kind of fluorescence is produced. In other words, because of this, these substances shine. So it was said that these rays must have been further modified. Thus we are dealing here with quite a number of different kinds of rays. The rays that came directly from the negative pole proved to be capable of modification by all sorts of other means. Then the attempt was made to find substances that were believed to make this change happen very strongly, in other words, that very strongly transformed these rays into something else, for example into fluorescent rays. And this is how we arrived at the fact that we can have substances like uranium salts, which don't always need to be radiated at all, but which under certain circumstances emit these rays themselves. That is, intrinsically they can emit such rays. And prominent among these substances were the ones known as radium compounds.

Certain of these substances have highly peculiar characteristics. First they radiate certain lines of force, which can be handled in a curious way. When we have radiation of this kind from a radium compound—the compound is in a lead container, and here we have the radiation—then we can investigate the radiation with a magnet. We find that something separates from the radiation and can be strongly deflected this way with the magnet so that it takes on this form [Figure 9c]. Another part doesn't bend and continues on in this direction. Yet another part is deflected in the opposite direction. In other words, the radiation contains three kinds of things. Finally, there weren't even enough names to describe it, so they called the rays that can be deflected to the right "beta rays," the ones that follow the straight line "gamma rays," and the ones deflected in the opposite direction "alpha rays." By placing a magnet to the side of the radiation, we can study the deflection, make certain calculations, and thereby determine the speed of the radiation. The result is interesting: the beta rays travel at about nine-tenths the speed of light and the alpha rays at about one-tenth the speed of light.[8] Thus we have certain explosions of force, so to speak, which we can separate and analyze, and which then show conspicuous differences in speed.

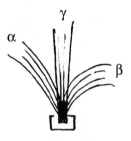

Figure 9c

At the beginning of these reflections we tried to grasp the formula $v = d/t$ purely spiritually, and we said that in space velocity is the real thing; it is the velocity that gives us the right to speak of something real. Here you see how what explodes out of here is mainly characterized by the fact that we are dealing with velocities that are acting upon each other with various strengths. Think for a moment about what it means that in the same cylinder of force radiating out from here, there is something that wants to move nine times as fast as the other; in other words one moving force that wants to hold back is asserting itself against the other force that wants to go nine times as fast. Now I want you to take a brief look at something that only we anthroposophists have the right not to regard as lunacy. Please recall how very often we had to talk about the fact that in the greatest world events we can comprehend, differences in speed are the essential element. How do phenomena of the greatest importance come into play in the present? By means of the fact that there is an interplay of the normal, the luciferic, and the ahrimanic influences,[9] that there are differences in speed in the spiritual streams the universe is subject to. The path that physics has opened up in recent times forces it to go into differences in velocity in a sense that, for the moment quite unconsciously, is very similar to the way in which spiritual science has to assert their significance for the most comprehensive agents of the world.

However, that doesn't exhaust everything that radiates out of this piece of radium. Another thing radiates out whose effects can in turn be proven. This reveals itself in its effects as something that radiates as an emanation of the substance radium,[10] but then gradually no longer manifests itself as radium, but rather as helium, for example, which is a completely different substance. Thus radium not only emits what is in it in the form of agents, but also gives up its own substance and becomes something quite different in the process. This no longer has much to do with the conservation of matter, but with the metamorphosis of matter.

Today I have presented you with phenomena that all take place in an area we could call the electrical realm. These phenomena all have something in common: they behave toward us in a way that is quite different from the way sound, light, and even heat phenomena behave. We float, so to speak, in light, sound, and heat in the way we have described in the preceding reflections. We can't say that so easily about electrical phenomena, for we don't perceive electricity as something that is as specific as light. Even when electricity is forced to expose itself to us, we perceive it by means of a light phenomenon. This has long since led us to say that electricity doesn't have a sense associated with it in human beings. Light has the eye as sense in the human being, sound has the ear, and for heat a kind of heat sense is construed. It is said that there isn't anything similar for electricity, which is perceived in a mediated way. However, we simply can't go beyond this characteristic of mediated perception unless we progress to a scientific examination of nature like the one we have at least inaugurated here. When we expose ourselves to light, we do it in such a way that we float in the element of light and take part in it ourselves, at least partially, with our consciousness. That is equally the case with heat and with sound. We can't say that about electricity.

But now I'll ask you to recall that I have always shown you that we human beings are actually, roughly speaking, double beings, in reality actually threefold beings: thought beings, feeling beings, and will beings. And I have shown over and over that we are actually awake only in our thinking, that we dream in our feelings, and that we sleep in the processes of our will, even when we are awake. We don't experience the processes of our will directly. We sleep through that which in essence is will. In these discussions I have pointed out to you that when we pass from the merely countable, from movement and time and space, to something that is not merely kinematic—where we write m = mass in physical formulas— we must realize that this corresponds to a transition of our consciousness into a state of sleep. If you take an unbiased look at this organization of the human being, you can say to yourself that to a high degree, the experience of light, sound, and heat falls into the realm we comprehend with our sensory/conceptual life. This is especially strong in the case of light phenomena. Thus simply because we are studying the human being in an unbiased way, those experiences reveal themselves as related to our conscious soul forces. When we progress to mass itself, to the material world, we are approaching something that is related to the forces that develop in us when we sleep.

We travel exactly the same path when we descend from the realm of light, sound, and heat to the realm of electrical phenomena. We don't experience our will phenomena directly; rather we experience what we can imagine of them. We don't experience the electrical phenomena of nature directly, but what they deliver up into the realm of light, sound, heat, etc. We enter the same lower world when we sleep that we enter in ourselves when we descend from our thinking, conscious life into our will life. Whereas everything that is light, sound, and heat is related to our conscious life, everything that takes place in the realm of electricity and magnetism is intimately related to our unconscious will life. And the appearance of physiological electricity in certain lower animals is only a symptom expressing itself in a particular place in nature of an otherwise imperceptible but general phenomenon: everywhere that the will acts through the metabolism, something similar to external electrical and magnetic phenomena is at work.

By descending along complicated paths, which we were only able to sketch out roughly today, we are actually descending into the realm of electrical phenomena, into that same realm we have to descend into even to arrive just at mass. What are we doing when we study electricity and magnetism? We are studying concrete matter. You descend into matter when you study electricity and magnetism! And what an English philosopher[11] said is very true: In earlier times people imagined in the most various ways that matter was the basis of electricity. Now we have to accept that what we believe to be matter is actually none other than fluid electricity. Previously, we atomized matter. Now we think that electrons move through space and have similar characteristics to those we used to attribute to matter.

Although we don't admit it, we have taken the first step toward overcoming matter and toward acknowledging that we are descending in the realm of nature when we make the transition from light, sound, and heat phenomena to electrical phenomena; we are descending to something that is related to those phenomena in the same way that our will is related to our thinking life. I would like you to take that to heart as the sum of our study today. I want to speak to you chiefly about things you won't find in books. What I do present to you from books, I only say as a basis for the other material.

Tenth Lecture

TODAY, TO CONCLUDE for the time being these few improvised hours of reflections on the natural sciences, I would like to give you a few guidelines that can be useful in creating your own nature studies based on characteristic facts you can demonstrate experimentally for yourself. In the natural sciences it is very important for teachers to find their way to the right forms of studying and thinking about what nature has to offer. In this regard, I was trying to show you yesterday that since the 1890s the course of physical science has been such that materialism has been turned on its head. It is this aspect that you should emphasize most.

We have seen that an era that believed it already had ironclad proof for the universality of wave phenomena was followed by an era that couldn't possibly hold on to the old wave or undulation hypothesis. The last three decades in physics have been as revolutionary as we could possibly conceive anything in a given field to be. For, under the press of the facts that came to light, physics has lost nothing less than its very concept of matter in the old form. We have seen that the phenomena of light were brought into a close relationship with electromagnetic phenomena and away from the old way of looking at things, and that the phenomena of electricity passing through rarefied air or gas tubes led us to see something akin to electrical emissions in the light emissions themselves.

I am not saying that this is correct, but it did happen. And we achieved that by eavesdropping, so to speak, on the course of the electrical current, which, in leaving the wire and jumping to a far distant pole, can't hide what it contains in the matter that it goes through. Before it had always been shut up in wires and could hardly be studied except according to Ohm's law. Because of this, something very complicated came to light. We saw yesterday that first the so-called cathode rays were discovered, which emanate from the negative pole of Hittorf's electron tubes and pass through a space with rarefied air—those were the phenomena I demonstrated for you, of course. Because these cathode rays could be diverted by magnetic forces, they revealed a kinship to something we normally regard as material. On the other hand, they are also related to something that we perceive in rays. This can be shown particularly vividly by carrying out experiments by which these rays that come somehow from the negative pole are captured, as light would be, on a screen or some other object. Light throws shadows, and radiation of this kind also throws shadows. Naturally, precisely because of this, its relationship to the ordinary material element is established, because, if you imagine that we are bombarding from this point, then the bombs don't go through the barrier, and what is behind it remains untouched (as we of course saw yesterday, in accordance with Crookes's ideas about what is happening with the cathode rays). We can illustrate this particularly well by capturing the cathode rays as in Crookes's experiment.

We will produce the electrical current here, then conduct it through this tube containing a partial vacuum, which has its cathode, the negative pole, here, and its anode, the positive pole, here. By driving the electricity through this tube, we get what are called cathode rays, which we then capture with the attached screen that is shaped like a diagonal cross. We have them strike it, and you will see that something like the shadow of this diagonal cross becomes visible on the other side, which tells you that the cross blocks the rays. Please consider this carefully: the diagonal cross is inside here, the cathode rays go this way, are blocked by the cross sitting here, and the shadow becomes visible on the back wall. Now I will draw this shadow that becomes visible into the field of a magnet, and I ask you to observe the shadow. You will find that it is influenced by the magnetic field. Do you see? Just as I can attract any other simple object, say, of iron, what has emerged here as a kind of shadow behaves like external matter. In other words, it exhibits the behavior of matter.

Thus, on the one hand, we have a type of rays here that, for Crookes, could be traced back to radiant matter—an aggregate state that isn't solid, liquid, or gaseous, but a finer state—and that shows us that in its flow electricity as a whole behaves like simple matter. We have focused our attention on the current of flowing electricity, and what we see there reveals itself to be like the effects we see within matter.

Now, because it wasn't possible yesterday, I also want to show you how the rays that come from the other pole arise, the ones that I characterized yesterday as anode rays. You see the difference here between the rays that are coming from the cathode, which are going in this direction and shimmering with a violet-like light, and the anode rays, which are coming toward them at a much slower speed and giving off a greenish light. I also want to show you the type of rays that arise by means of this device here; they will reveal themselves to you in that the glass becomes fluorescent when we pass the electrical current through it. Here we will get the type of rays we can otherwise make visible by passing them through a screen made of barium platinocyanide, which have the characteristic of making the glass very strongly fluorescent. You see the glass—I want you to focus your attention mainly on this—in a very strongly greenish-yellow fluorescent light. The rays that appear in such a strongly fluorescent light are none other than the X rays I mentioned to you yesterday. So we've taken note of this kind here, too.

Now, as I've told you, in the pursuit of these processes it turned out that certain of these entities that were regarded as substances emit whole bundles of rays, initially at least three different kinds, which we classified as alpha, beta, and gamma rays and which reveal characteristics that clearly distinguish one from the other. These substances, called radium and so forth, emit yet a fourth substance, which is the element itself.[1] The radium sacrifices itself, so to speak, and, after being emitted, transforms in such a way that while the radium is streaming out it changes into helium, thus becoming something quite different. In other words, we are not dealing with a constant substance, but with a metamorphosis of the phenomena.

To follow up on these things, I would just like to develop an aspect that can become a path for you into these phenomena, indeed the path into natural phenomena in general. You see, the chief reason that the thinking of nineteenth-century physics became sick is that the inner activity by which people sought to pursue natural phenomena was not agile enough in the human being and, above all, was not yet capable of entering into the facts of the external world itself. We could see color emerge in and under light, but we didn't rise to receive color into our imagination, into our thinking. It was no longer possible to *think* colors, and we replaced the colors we couldn't think with something we could think, something that is merely kinematic—the calculable vibrations of an unknown ether. This ether, however, is tricky, because it doesn't present itself whenever you seek it. And all these experiments actually showed that flowing electricity does indeed reveal itself as something that exists as a phenomenal form in the external world, but that the ether doesn't want to present itself at all. Now it just wasn't given to the thinking of the nineteenth century to penetrate into the phenomena themselves. However, from this moment onward that is exactly what will be so necessary for physics—to go deep into the phenomena themselves with the human imagination. To do that, certain paths have to be opened up, particularly for the study of physical phenomena.

We would like to say that by approaching closer to the human being, the objective powers have actually already forced our thinking to become more agile, but we could say that this has happened from the wrong starting point. What we regarded as a sure thing, something that we relied on the most was of course what we were able to explain so beautifully with calculation and with geometry, in other words with the arrangement of lines, of planes, and of bodies in space. But the phenomena in Hittorf's electron tubes force us to approach the facts more closely, because calculation indeed actually fails if we try to apply it in such an abstract form as the earlier wave theory did.

First I would like to talk to you about the starting point from which something like a compulsion came to make arithmetical and geometrical thinking become agile. Geometry is very old. The way that we imagine the laws of lines, triangles, rectangles, etc., on the basis of geometry is an ancient inheritance we have applied to the external phenomena that nature presents us. However, in the face of nineteenth-century thought this kind of geometry began to totter. It happened in the following way. Transport yourself once again back onto the old school bench. You were taught—and it goes without saying that our dear Waldorf School teachers teach it too, and have to teach it—that when we have a triangle and calculate the three angles, those three angles together are equal to a straight angle, or 180 degrees. Naturally we feel compelled—and have to feel compelled—to give the students some kind of proof for the fact that these three angles together are 180 degrees. We do this by drawing a parallel here to the base of the triangle, saying: the same angle that is here as α appears here as α^1; α and α^1 are alternate interior angles. They are equal. Thus I can simply put this angle over here [Figure 10a]. Likewise, I can put angle β over here and have the same thing. Now, angle γ stays where it is, and if $\gamma = \gamma$ and $\alpha^1 = \alpha$, and $\beta^1 = \beta$, and $\alpha^1 + \beta^1 + \gamma$ together make a straight angle, then $\alpha + \beta + \gamma$ together also have to make a straight angle. I can prove this clearly and concretely. There can't be anything clearer or more concrete, you might say.

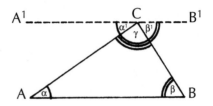

Figure 10a

However, the assumption that we make here when we prove this is that this upper line A^1B^1 is parallel to AB, because only then am I able to carry out the proof. But in all of Euclidean geometry now there is no way of proving that two lines are parallel, that is, that they intersect only at an infinite distance; in other words that they don't intersect at all. This looks as if they are parallel only as long as I stay with imaginary space. Nothing guarantees that this is also the case with real space. And if I assume only one thing—that these two lines do not intersect only at an infinite distance, but in reality intersect earlier—then my whole proof that the angles of the triangle equal 180 degrees is wrecked. Indeed, in the space that I construe for myself in my thoughts and with which normal geometry is concerned, the sum of the angles of the triangle is 180 degrees.

But as soon as I contemplate a real space, it may be different. Then I would conclude that the sum of the angles of the triangle isn't 180 degrees at all, but perhaps greater. In other words, besides the normal geometry that stems from Euclid, there are other possible geometries, for which the sum of the angles of a triangle is not 180 degrees at all.

Especially since Lobachevsky,[2] nineteenth-century thinking was much occupied with debates in this direction, and in the aftermath the question had to arise: Can the processes of the real world that we follow with our senses actually be comprehended, completely comprehended, with those concepts we generate as geometric concepts in an imagined space? The imagined space is without a doubt imaginary. We can indeed foster the beautiful notion that what happens outside of us out there partially coincides with what we cook up about it, but nothing guarantees us that what happens out there works in such a way that we can grasp it completely with the Euclidean geometry we have thought up. It could easily be the case that things out there happen according to a quite different geometry and that it is only we in our interpretation who translate it into Euclidean geometry and its formulas. Only the facts themselves can tell us the truth about this.

That means that, initially, if we accept only what the natural sciences currently have at their disposal, there is no possibility for us to decide how our geometric or, more generally, our kinematic concepts relate to what we see out there in nature. We calculate and draw natural phenomena, as long as they are physical. However, whether we are only superficially   drawing just any outward thing or penetrating into something natural can't be determined for the time being. Once we begin to think really thoroughly, particularly in physics, we will end up at a terrible dead end and see we aren't getting anywhere. And we will get somewhere only if we first educate ourselves about the

origin of our kinematic concepts, of our concepts about counting, about geometry, and also about movement purely as movement, but not of our concepts about forces. Just where do all
these kinematic concepts come from? We may normally believe
that they arise from the same source as the concepts we develop
when we involve ourselves in the outer facts of nature, and
work on them rationally. We see with our eyes and hear with
our ears. What we perceive with the senses we process with our
intellects in a primitive way initially, without counting it, without drawing it, without looking at its motion. We are guided
by completely different conceptual categories. Our intellects
are active in the presence of sense phenomena. However, when
we start to apply so-called scientific concepts of geometry,
arithmetic, algebra, or motion to what happens externally, we
are doing something quite different. We are applying concepts
that we haven't gotten from the outer world, but have con-
cocted from our inner selves. Just where do these concepts
come from? That is the cardinal question. These concepts
absolutely do not come from the intelligence that we apply
when we process sense perceptions. In fact, they come from the
intelligent part of our will. We make them with our will structure, with the will element of our soul.

There is a tremendous difference between geometric, arithmetic, and motion concepts and all the other concepts of our
intelligence. We gain the other concepts from our experiences
of the outer world. Geometric and arithmetic concepts arise
from our unconscious selves, out of the part that is will, which
has its external organ in the metabolism. Most prominently, for
example, the geometric concepts arise from it; they come from
the unconscious in the human being. And when you apply
these geometric concepts (and arithmetic and algebraic
concepts as well) to light phenomena or sound phenomena, in
your cognitive process you are linking what arises within you to

your external perceptions. In this process you remain unconscious of the whole origin of the geometry that you have used. The whole origin remains unconscious. And you develop such theories as the wave theory (of course it makes no difference whether you develop this one or Newton's emission theory) by uniting what arises in your unconscious self with what represents your conscious daily life, the sound phenomena, and so forth. You permeate one with the other. But these things don't belong together. They belong together as little as your imaginative capacity belongs together with external things you perceive in a kind of half sleep.

In anthroposophical lectures I've often given you examples of how human dreams employ symbols. A person dreams he is a student standing at the door of a lecture hall with another student. They have an argument, which becomes violent. They challenge each other to a duel—this is all a dream. He dreams that he goes out into the woods. The duel is arranged. The dreamer even dreams that he fires. At that moment, he wakes up—the chair has fallen over. That was the jolt that is played out in the dream. Imagination became connected with an external phenomenon in a purely symbolic way, not in a form that suited the object. The kinematic concepts that you fetch up out of the unconscious part of your being are connected in a similar way to light phenomena. You draw rays of light geometrically. What you are doing in this case has no more validity than what is expressed in the dream when you imagine symbolically such objective facts as the bang of the falling chair. This whole way of processing the optical, acoustical, and heat phenomena of the external world by geometric, arithmetic, and motion concepts is in truth a waking dream, albeit a very sober one. Until we recognize that it is a waking dream, we won't deal with natural science in such a way that it supplies us with realities. What we believe to be an exact science is modern humankind's dream of nature.

However, if you descend from light phenomena, from sound phenomena through heat phenomena into the area that you enter with radiation phenomena, which are simply a special chapter of the theory of electricity, you are connecting with something external in nature that is the equivalent of the human will. The will area in the human being is equivalent to the area in which the cathode, anode, and X rays, the alpha, beta, and gamma rays, etc., are operative. From this will area in the human being arises what we have in our mathematics, in our geometry, and in our concepts about motion. Only at this point do we enter related areas in nature and in the human being. Nevertheless human thought in these areas is not yet advanced enough to really penetrate them with thinking. Modern people can dream, thinking up wave theories, but they can't yet grasp mathematically this area of phenomena to the extent that it is related to the area of human will from which geometry and arithmetic arise. For that to happen, our arithmetic, algebraic, geometric imagination itself has to become more thoroughly permeated with reality, and this is precisely the path that physical science must take.

These days, when you talk to physicists who were educated in the era when the wave theory was in flower, many of them are quite uneasy about these more recent phenomena, for they are causing arithmetic concepts to break down in every possible way. Of course, in recent times, because completely legitimate arithmetic and geometric procedures didn't work anymore, we have found another way of doing things. A statistical method has been introduced that allows us to connect empirical numerical relationships more closely to the external empirical facts and to work with calculations of probability. This allows us to say that a given regularity can be calculated that lasts for a certain series; then a point comes when it doesn't work that way anymore.

Such things show how, especially in the course of the development of more recent physics, we may indeed lose a thought, but, in the very process of losing it, arrive at reality. Thus, for example, it is easily conceivable, given certain rigid ideas about the nature of a heated gas or heated air and its behavior vis-à-vis the environment under certain circumstances, that someone could have proven with mathematical certainty that the air could never be liquefied. Yet it *was* liquefied, because at a given point it was shown that certain ideas that account for the laws of a series have no validity at the end of a series. I could cite many examples of this kind. Such examples show how today, especially in physics, reality often forces people to admit that with their thinking, with their concepts, they are no longer immersing themselves fully in reality. They have to start the whole thing from a different point.

To start from this different point, it is necessary to feel the relationship between everything that arises from the human will—and that is where kinematics comes from—and what is separate from us and approaches us externally in such a way that it announces its presence only in phenomena of the other pole. Everything that passes through those tubes uses light to announce itself, and so forth. But the flowing electricity is not perceivable by itself. That's why people say that if we had a sixth sense for electricity, we would be able to perceive it directly too. That's nonsense, of course, for it is only when we ascend to intuition,[3] which has its basis in the will, that we come into the region where electricity lives and works, even for the outer world. Along with this, however, we also notice that, in this last area we have examined, we have the inverse of what we have in the case of sound. With sound the peculiar thing is that because people are simply immersed in the world of sound, as I described it, they live into sound itself only with the soul, whereas with the body they only live into something that sucks at the real essence of sound, in the sense that I have looked at things in these last few days. You will recall the analogy to the globe from which the air had been pumped out—it sucks at the sound! With sound I am within, in the essence of the spirit, and what is observed by the physicist, who naturally is unable to observe things of the spirit, of the soul, is the external "material" phenomenon that is the parallel of motion, of the wave.

If I go to the electrical phenomena of the last area we have examined, I don't have only the objective so-called material world outside of me, but also what otherwise lives within me, in spirit and soul, as sound. In essence, sound is also present externally, but I am bound up with this thing that is outside. With these electrical phenomena, on the other hand, I have, in the same external sphere in which in the case of sound there are only material sound waves, something that with sound can be perceived only by the soul. With electrical phenomena I have to perceive physically the same thing that in the case of sound I can perceive only with the soul.

In terms of the relationship of the human being to the outer world, sound perceptions and perceptions of electrical phenomena, for example, are at diametrically opposite poles. When you perceive sound, you divide yourself into a human duality, so to speak. You float in the wave element, the undulatory element that can be verified externally. You become aware that there is something more within it than the merely material. You are forced to become inwardly active to comprehend the sound. You become aware of the undulation, the vibrations, with your body, your ordinary body, which I'll draw schematically here [Figure 10b]. You retract your ether and astral bodies together into yourself, which then occupy only part of your space, and experience in the inwardly concentrated etheric-astral body of your being what you are supposed to experience in the sound. When you confront electrical phenomena as a human being, at first you perceive nothing at all in the way of vibrations and the like. But you feel compelled to expand what you had previously concentrated [Figure 10c]. You push your etheric body and astral body out past your surface, and by enlarging them, perceive these electrical phenomena.

Figure 10b

Figure 10c

Without progressing to the human spirit and soul, we will never be able to gain an understanding of natural phenomena that corresponds to truth or reality. We have to imagine more and more clearly how sound and light phenomena are related to our conscious conceptual element; electrical and magnetic phenomena, on the other hand, are related to our subconscious will element; and heat is located between them. Just as feeling is located between thinking and willing, the external heat of nature is located between light and sound on the one hand and electricity and magnetism on the other. Thus the structure for examining natural phenomena must increasingly become the study of the light and sound element, on the one hand, and of the diametrically opposed electrical-magnetic element on the other. And this can happen if we pursue the Goethean theory of color. Just as in the spiritual realm we distinguish between luciferic/light-bearing and ahrimanic/electromagnetic,[4] we also have to study the structure of natural phenomena in this way. And the phenomena of heat that we encounter are located in neutral territory between the two.

With that I have indicated a kind of path for you—guidelines, in which I have tried to summarize for the moment what I was able to present to you in these few improvised hours. Given the haste with which the whole thing had to be put together, it goes without saying that it didn't go beyond my intentions. I could only give you a few ideas, which I hope to be able to expand on in the near future.[5] I believe, however, that what I've given you here will help you, and can especially help the teachers at the Waldorf School when you teach the children concepts about the natural sciences. You have to see to it that you don't teach the children directly, I mean to say, in a fanatical way, so that they go right out into the world and say, "All university professors are asses." Instead, with these things what is important is that the realities be developed in an appropriate way. We shouldn't confuse our children, but at least we can succeed in not mixing too many impossible concepts into our lessons, concepts based solely on the belief that the dream image we construct about nature has an actual external reality. If you let yourselves be pervaded by a certain scientific cast of mind like the one that pervades what I've presented to you in the last few hours, for example, then that can serve you in the way you talk to the children about natural phenomena.

I believe that you can also get a good deal out of it in terms of methods. Although I would have preferred not to rush through these phenomena in the way I had to, you will nevertheless have gathered that we can connect what is observed externally in an experiment with the concepts that are called forth about things so that we don't simply gape at things, but reflect on them. And if you set up your lessons in such a way that the experiment makes the children think and you discuss the experiment with them rationally, then especially in the science lesson you will develop a method that will make the natural sciences fruitful for the children entrusted to you. Thereby I believe I've added something by way of example to what I said in the pedagogical course when instruction began at the Waldorf School. [6]

On the one hand, I believe that, by setting up these courses, we have once again done something that can help our Waldorf School thrive, so that it can really develop—and it can do just that in the wake of its very commendable start. This should be the beginning of work that draws on something new for the development of the human race. If we let this consciousness permeate us—that there is simply so much that is breaking down in what was built up thus far in the development of the human race and that something newly formed has to take its place—then we will have exactly the right consciousness for this Waldorf School. Particularly in physics, quite a few concepts have proven to be extraordinarily fragile, and this fact has a much stronger connection than we might think with the misery of our times. Isn't it true that when people think in sociological terms, we notice right away how their thinking has gone awry? Most don't even notice that, but we are able to notice it because sociological ideas have an effect on human social order. But we don't really form any satisfactory idea of how deeply the concepts of physical science affect all of human life, so we are ignorant of the damage that has, in truth, been caused by the often horrific ideas of modern physics.

Indeed, I have often cited how Hermann Grimm,[7] who for his own part viewed the ideas of the natural sciences only as an outsider, stated with a certain justification that future generations would hardly be able to comprehend that there was once such a crazy world that explained the development of the earth and the whole solar system on the basis of the theories of Kant and Laplace. Someday in the future, it won't be easy to grasp this scientific insanity. However, there is a great deal today in our ideas about inorganic nature that is like Kant's and Laplace's theories. But how are people going to free themselves of Kantian-Königsberger thought and the like if they want to move forward to far-reaching, healthy concepts?

We hear of strange things in which we can see that the wrong on one hand is linked to the wrong on the other. It can make your skin crawl to hear such things as the following. Recently I was shown—coincidentally, as they say—a copy of a lecture that a German university professor, who even declares himself in the lecture to be a Kantian, gave on May 1, 1918, at a university in a Baltic state about the relationship between physics and technology. Please take note of the date: the first of May 1918. The man, who is a learned physicist, expresses his ideal at the close of the lecture, saying more or less that the course of this war has clearly shown us that we haven't been able to forge to an adequate degree the union between the scientific work of the university laboratories and the military. In the future, so that the human race can make appropriate progress, a much closer bond must be made between the heads of the military and what is happening at the universities, for everything coming from science that can make mobilization especially powerful must in the future be included in mobilization issues. At the beginning of this war we suffered from the fact that this intimate bond had not yet been joined, a bond that in the future should therefore lead from the institutes of experimental science into the general staffs.

My dear friends, the human race must change its ideas, and it must change them in many areas. If we can decide to change them in such an area as physics, it will be easier for us to change our ideas in other areas, too. Those physicists who go on thinking in the old way, however, won't ever be far removed from this nice little coalition between the institute of experimental science and the general staffs. A great deal has to change. May the Waldorf School always be a place where the changes that are to come can take root! With this wish, for the time being, I would like to bring these discussions to a close.

Discussion Statement[1]

DORNACH, AUGUST 8, 1921

AT THE CONCLUSION of her very noteworthy remarks, Dr. Rabel said that I had once commented that these more recent experiments could actually serve to confirm the Goethean theory of color. At that time Dr. Rabel was so kind as to give me one of her treatises[2] that are along exactly the same lines, and I said that the facts coming out of modern physics are indeed of such a nature that they would gradually have to lead to a confirmation of Goethean color theory.

Today there is unfortunately no possibility of going into all the pros and cons of Goethean color theory and, let's say, anti Goethean color theory. As the matter stands, for the time being the concepts of physics that are the norm today are based on theoretical premises of such a nature that what I once heard from a physicist with whom I had a conversation about Goethean color theory[3] is indeed correct. He said simply and honestly—as I can only verify in unmistakable terms—that a physicist of the present day can't make heads or tails of Goethean color theory! And that is actually quite true.

We must not forget that certain things have to be overcome before Goethean color theory is to be taken seriously by physics. Isn't it true that the physicists of today feel bound to investigate what they call light in such a way that, as far as possible, the "subjective" plays no role in the field of investigation? To a certain extent, the experience they have of light phenomena serves at most to make them more attentive to the fact that something is happening there. However, what the physicists want to include in their interpretations of light phenomena—which they also extend to color phenomena—is an entity that is completely independent of subjective experience.

Goethe starts out from completely different premises for his whole way of thinking. Therefore I still consider what I said in 1893 in a lecture in Frankfurt am Main about Goethe's view of nature to be right in a certain sense.[4] I said it was possible to talk about Goethe's statements in the field of morphology, and even at that time I gave a lecture about that, because the ideas that Goethe had about metamorphosis and about the origin of species in connection with metamorphosis already coincided to a certain extent with those that were coming, although in a completely different way, from the Darwin-Haeckel point of view. Thus, in a certain sense at least, there was already a field in which the points of view overlapped. However, with Goethe's color theory—which doesn't aspire to be a theory of optics, by the way—that is not yet the case at all. Therefore, on the basis of anthroposophy, let's say, it is indeed possible to talk about Goethean color theory. A conversation is certainly possible there, but a discussion about what physicists have to say about color, about what they derive from the physical basis of their science, is still going to be completely fruitless even today. For that to happen, certain basic concepts that were implicit with Goethe and were the starting point for his theory of color would still have to be explained so that they could really be made the basis.

Therefore I consider everything I have said in my books about Goethean color theory to have been put out into the world for the time being without having any pretension of beginning a productive discussion with the ideas of physics, which, while not contradictory, come from a completely different angle. Now, however, you can be quite certain—and the previous speaker already said a great deal about this—that Goethe would see a confirmation of his basic point of view in all of the phenomena Dr. Rabel so kindly took up today. And that is a proposition I would absolutely defend.

If, speaking with Goethe, we say that one side of the spectrum—in other words, what has been called long-wave radiation as opposed to short-wave radiation—is in the relationship of a polarity, this corresponds to the facts of the matter from one angle, but not completely. Polarity is a very abstract relationship, which we can simply apply to various opposites and thus to this one too. Only that isn't really the point at all with Goethe here.... [a gap in the transcription] ...

However much we may believe we can exclude mistakes by some kind of experimental arrangement whereby we make the bundle of rays narrower and narrower, such that the entire thickness of the bundle of rays—which, by the way, isn't my expression, but I may employ it here legitimately—is finally canceled, and then speak of one "ray," in the end there is no difference in reality. Whether we take a broad bundle or a narrow one, in principle it doesn't make any difference. Goethe, however, did indicate a difference in principle when he conducted experiments with the small opening.

In the prism we can't exclude what modern physics would like to exclude, for of course we can't insert a so-called "ray with a thickness of zero" into the experimental field somehow. However, it is possible to observe the sharp edge between the dark and bright areas. Indeed there you do have the sharp edge! When we speak of the sharp edge, to a certain extent we are getting out of the Goethean experiment exactly what recent physics would like. Goethe worked with the edge and not with the bundle of rays. That's what is important. This legitimate demand we make for the ideal is fulfilled by the fact that Goethe works with the edge and not with a ray, in other words, or a bundle of rays. And Goethe starts out from what happens at the edge and tries to base the setups of his experiments on that; however, if they were to be carried out in the Goethean sense today, they would have to be carried out quite differently.

I hope that especially in this connection we will be able to undertake basic experiments in our physical research institute in Stuttgart and that by doing so we will eliminate to an extent what Dr. Schmiedel[5] called "veiling" and really learn how to work in an exact way with the edges. Only then will we be able to understand the spectrum as a phenomenon in which the edge phenomena are treated as archetypal or *Ur*-phenomena.

Now, however, if we work in this way with the edge, then we get what Dr. Schmiedel called the polar relationship between one part of the so-called spectrum and the other.

Thus in the Goethean sense the expression "polarity" is much too abstractly applied here. You can use it as an expression for all sorts of different natural phenomena. For lack of time I can't go into the details this evening, but, by constantly trying different experiments, Goethe came to accept a basic opposition between red nature and blue nature.[6] It is important to note that Goethe doesn't speak of red and blue light— that is inconsistent with his sense—but of red and blue *nature*. Light simply cannot be differentiated, and any kinds of differentiation that may appear are phenomena *in and under* light. Recent physics makes it possible for us to emphasize that Goethe sets up the entity he calls light against the entity of darkness, not as nothingness, but as a real entity. Now I can only describe in a few short words what is actually a fairly complicated concept of Goethe's. In the red part of the color tones as well in the blue part we are dealing not with a mixture, but with an interaction of light and darkness, such that in the red part, the color red is the result of the activity of light in the darkness. If we are dealing with red, let's say a red field, we are dealing with light that is active in the darkness whereas, if we are dealing with the blue side, then we are dealing with the activity of darkness in the light. That is the exact way to express the polarity.

I'll admit that a modern physicist naturally can't make much sense of this concept. However, for Goethe red is the activity of light in darkness, and blue is the activity of darkness in light. We can call that a polarity. It *is* a polarity. Goethe carries this through with physical color, spectral color actually, and also with chemical color. He is well aware that he is running into uncertainties everywhere, because of course he is unable to apply this general principle in specific instances. But if we follow this thought that I have just hinted at in passing, then everywhere that colors occur—where there are color phenomena, in other words—we have something qualitative. And now we have come to the point where someday the decision in this regard will be made.

Nowadays it is still the case, if I may say so, that we experience an abundance of phenomena. Even today you have been generously presented with an abundance of phenomena, about which we would have to give whole series of lectures to show how they actually fit into Goethean color theory and into the whole field of the natural sciences. However, the phenomena we are experiencing nowadays call for corrections of a completely different sort than those given by the theoretical considerations of relativity theory, etc., concerning ideas about the speed of light. As Dr. Rabel herself just emphasized, what we are experiencing right now is that physicists feel constrained to turn to Newton's emission theory once again, although in greatly modified form. Of course, there is a big difference between the Newtonian theory, which was drawn from relatively simple phenomena, and the view of the current era, for the current view is based primarily on the fact that, with the usual concepts of wave theory, we can't imagine how the following phenomenon, for example, is possible.

If we cause ultraviolet light to strike a metal, electrons are reflected, which can then be studied. These electrons reveal a certain strength, which is not dependent on the distance of the source of the ultraviolet light from the metal. You can place the source far away and still get the same voltage. Now, assuming that the strength of the light remains the same, the intensity would naturally have to decrease progressively in proportion to the distance. However, this is not the case for these electrons reflected by the metal. We see that their strength doesn't decrease at all in proportion to the increase in distance. Rather it is dependent only on the color. The strength of the color is the same when the source is nearby as when it is at a greater distance. This leads us first of all to the conclusion that in general we have to think quite differently about this thing we call light. These days we help ourselves get around this fact by referring to quantum theory, which states that light isn't something continuous that spreads, like gravitation, for example, but instead spreads atomistically. If it spreads atomistically, then we have a given quantum at a given place, where it acts. That isn't the issue.... The quantum simply can only be at one place. If it is there at all, it triggers the electron effects.

These things have led us back to emission theory once again. While Newton imagined that substances, entities, spread out in a measurable way and are such that the intensity decreases with the square of the distance, we now replace these substances with the spreading out of electromagnetic fields, which, however, do actually pass through space in the way described by quantum theory. In other words, we are actually dealing with the emission of electromagnetic fields, whereas the wave theory, which was predominant at the time when I was young, deals only with the simple propagation of motion. Thus in the latter case nothing actually radiates in space; only the motion continues. These ideas about what is objectively present are in continuous flux right now, and the known experiments point in every case to what Dr. Rabel so rightly emphasized: that we can't get along with the mere supposition of wavelengths, that it comprises a kind of contradiction in terms. But that's exactly what we are dealing with here.

Basically, the situation is simply that for long periods of time we became accustomed to calculating solely in terms of wavelengths and so forth. The concept was extraordinarily simple. In general we calculated only objectively in terms of waves of certain wavelengths and vibrations of certain speeds, describing what lies in the spectrum from violet to red by saying that this is the range that makes an impression on the retina of the eye. Beyond red we have other vibrations that don't make any impression, but they are no different qualitatively, and it's the same on the other side of violet. Some rebelled against this. Some rejected it in an interesting way. For example, Eugen Dreher[7] conducted many experiments in the 1870s to prove that light, heat, and chemical entities are three completely, radically different things. And to a certain extent that could also definitely be proven. The current state of affairs particularly proves that the whole complex of issues is basically in flux. As soon as we arrive at what is actually known, which, apart from the subjective element, is summed up under the complex of "light phenomena" ... [gap in transcription] ...

The essential thing with Goethe is that he introduced the element that is forcing itself upon physics today. Of course he introduced it according to the inadequate state of physics at the end of the eighteenth century. Nevertheless he did introduce it.

Looking at this matter today, we say to ourselves that this is certainly all tremendously interesting. And I must confess that the whole treatment of wave theory was more interesting when

I was young because wave theory was developed to excess, and everything was really calculated quite exactly down to the smallest detail. These days young people are not plagued with this extravagant wave theory anymore. It's a somewhat different matter, whether, based on theoretical mechanics, we calculate the undulation with some kind of ether hypothesis or we start from the way electromagnetic fields work. All of that does indeed look a bit more uncertain. Today we don't have a need to calculate everything within light phenomena in such a straightforward way as was still done thirty to forty years ago. Naturally it is extraordinarily interesting to work out all these refinements, but they are the result of calculation, and the whole, decisive proof for this calculation is actually seen in the interference experiment. Nowadays the interference experiment is in need of a new explanation. Physics today is willing to admit that. And quantum theory really hasn't achieved much there.

The situation is as follows. We haven't gotten very far these days, but we increasingly see how certain very usable numbers that we have in the oscillation figures or wavelengths are all fine tools for calculations, but no one can actually say today that they have any basis in reality. I might say that if we state the wavelength for the so-called red rays and for the blue ones, we have a certain ratio between red and blue that expresses how one number is related to the other. Indeed today we can say that the ratios of the different numbers to each other are much more important than the absolute value of the individual wavelengths, which takes us from the quantitative into the qualitative. Today we are on our way to saying that we can't manage it only with wavelengths. We need something else.

But this "something else" is getting more and more similar to what Goethe was looking for along his own paths. This isn't clearly recognized yet, but those with an exact knowledge of these things can definitely recognize that physics is gradually leading in this direction. As I said, Goethe would definitely see the phenomena that were presented today as a confirmation of his point of view.

Naturally it is difficult to go into details because as of now the basis has not been created for them. I only want to go into the principles of the plant question, for example. I don't want to go into such things as whether we should be using an expression such as "absorbed" or not. If you take it merely as a description of what is known, then I have nothing against it. But we make the situation too simple for ourselves by saying that when a pane of glass is placed in the path of light and behind the glass is a red field, that all the other colors have been swallowed up by the glass, and only the red was allowed to pass through. Then we are replacing a perceived phenomenon with an explanation that comes completely out of the blue, for which there is no known reality. We can definitely stick with the phenomenon. That is good. However, let's look at how Goethe expressed it, still very imperfectly perhaps: the activity of light, of brightness, in the darkness is the basis of red; the activity of darkness in the brightness, in the light, is the basis of blue. What the basis of the nuances is, the shadings of green or orange, isn't important right now; I can't go into that. I can only indicate the basic phenomenon. Then, of course, you have what I indicated just now in a sketchy way. We are dealing here with darkness as a reality. We have to realize—there is a good deal of evidence for what I am going to say now, but even looking at the matter superficially we can come to an understanding of this—that this entity of darkness opposes the light in a certain way. Our subjective feeling tells us this, naturally, but so do the objective facts. We have to assume a polarity here if we don't want to stay with abstractions and instead go into the concrete facts.

Now, if you think about this polarity of light and dark, you will gradually become aware that with darkness it is impossible in a way to speak of the spreading of an entity the way we do with light. The experiments conducted up to now can't tell us anything about this. Just imagine that we describe light as a spreading out that takes place. Naturally there is more to it, but that is based on supersensible or partially supersensible observations, so let's take it just as a possibility, a hypothesis, for the time being. You can't then describe darkness as a spreading out that takes place. Instead you have to describe darkness as a kind of absorbing that takes place out of the infinite. You wouldn't be able to say in the case of a room furnished with black walls that a spreading out occurs, an emission, or the like. Rather an absorption takes place; there are absorptive effects that must have a cause because they must have a center. But for the time being it is the potential of absorptive effects that we are dealing with in this black room, speaking trivially, in contrast with the illuminated space we are dealing with in the case of spreading effects.

If you bear that in mind, the concept of color becomes more and more concrete; and in blue you have something that absorbs—this is actually only an approximation—while in red you have something that spreads out, and in green the neutralization, so to speak. Now, just think—here we have to go into a deeper layer of the imagination—if you look at the absorptive effects that are present in relation to plant life, then behind the colors you have an absorptive effect that stands in contrast to certain inner forces of the plant. Here you have something that plays a role within the whole configuration, in the whole organization of the plant.

Having grasped that, we also get ideas that are much more complex than if I say, "I place a pane of glass in the path of a light beam and get a red field behind it. Everything besides the red has been swallowed up." That takes us to something

completely different, to a completely different formulation of
the problem. The phenomenon before me demands that I
investigate the nature of the material placed in its path. If we
begin there, it leads us to a wholly different method of looking
at polarization phenomena, for example. In a roundabout way,
we arrive at a strict conception, which is what Dr. Rabel also
said. [To Dr. Rabel:] You named one English physicist, but
quite a few physicists have already called attention to this problem, that in the case of these phenomena we are not actually
dealing with something that points to the nature of light, but
with something that points to the nature of the matter that is
placed in opposition to the light, particularly the organic matter of plants. That is where this is leading us more and more—
to stop construing polarization as light. That's something that
worked wonderfully with the old, purely mechanical wave theory, but isn't valid in the same way in the current situation.

Now physicists aren't forced to see the occurrence of polarization phenomena in such a way that they merely build them
into light as constructs. Instead they observe an interaction of
light with matter such that the composition of the substance is
revealed by what appears there, even in the case of other phenomena that originate in such a way that we regard them as the
emission of electromagnetic waves. Looking at these things
today, it is much more interesting to observe how we are gradually weaning ourselves from a method that is based solely on
our being so completely accustomed to this mechanical view
with the ether, which some, of course, imagine as solid and
others as fluid.... [gap in transcription] ...

Indeed we have grown accustomed to certain ideas and
can't get away from them.... If we stick with the wave theory,
we have to assume that we have to find a different basis for
it.... And here we have to point out that Goethe was on his
way to investigating this basis. The whole wave theory, which

he had been familiar with for his whole life, didn't interest him. Rather he was interested in something I just touched on inadequately when I traced polarity back to the concrete.

We delve deeper into what Goethe wanted by taking his *Theory of Color* chapter by chapter, even ascending to the sensory-moral effects of the colors, where, to a certain extent, color disappears from the field of view, and, if I may say so, qualities of spirit, soul, and morality appear. We experience them in the place of red and blue when we are transported into the realm of the soul. And Goethe would say in this case that it is only then that we actually learn something about the essence of color—when color disappears and something wholly different appears.

What appears is the beginning of the paths to higher knowledge described by anthroposophically oriented spiritual science,[8] which lead us to abandon the separation of subject and object that has no place on a certain level of knowledge, and which instead lead the subject to live into the object. This is something we have to pay attention to. There can be no satisfactory theory of knowledge[9] if an absolute chasm lies between subject and object, but only if this subject/object classification is a temporary assumption, as has been described epistemologically. Modern physics in the way that, say, Blanc[10] defines it certainly has the goal of excluding the subjective completely and describing the phenomena as they occur in an objective field without any consideration of the human being. Louis Blanc says that physics should only look for what an inhabitant of Mars—even if organized in a completely different way—could also claim to be true of the objective world. And that is definitely right. But the question is, can we also find something in the human being that corresponds to the results of physics, which are sought purely in terms of measurement, number, and weight? Is there something in the human being that, with correspondingly higher knowledge, corresponds to that? And here we have to say, Yes, there is! We pass right through this region, which is then experienced, and which the modern physicist actually recovers only by a construction, a certain construction based on the phenomenon. Only the way this region looks is such that the substance forming its basis is no longer material, but spiritual in nature. We even gain the right to apply the formulas of physics in a certain form, just plugging in a different kind of substance. Newton thought that a kind of measurable matter is plugged into the equations, the formulas; in Huygens's wave theory, only the number of waves is plugged in; according to more recent theory, electromagnetic fields are plugged in.

Thus today, in terms of what is actually contained in the formulas, a certain liberality in the development of the theories is already the rule. Therefore we shouldn't resist so very much if spiritual science finds it necessary now to add spirit to these dancing, space-traveling equations. Neither what Newton wanted, nor what the thoroughly modern physicist wants— instead, just add spirit to them! Only first we just have to know what spirit is. That isn't based on any theory but on higher knowledge.

Thus I believe that more and more is being contributed to a true understanding of Goethe's theory of color because of the things that Dr. Rabel was kind enough to present today. Nevertheless I do not believe that it is possible yet to go into such questions as those posed, for example, by Dr. Stein,[11] because then we would have to go into the whole nature of electricity. And that touches on questions that can only be addressed—I don't say solved—in the realm of anthroposophy. That is because we begin then to arrive at concepts that overturn everything we are currently accustomed to recognizing theoretically about the physical world.

Even if we have gotten away from it somewhat these days, it hasn't been so long since we were calculating in terms of electrical currents and the like. Now, however, what we are actually dealing with in the case of electrical currents—what I'm going to say to you now is solely the result of higher knowledge—is not something that streams in. Rather in reality we are dealing with the fact, if I may indicate it schematically, that if we have a wire here that a so-called electrical current is flowing through, in reality we have a gap.

If I want to designate reality—I am speaking now of a degree of reality that of course many won't find valid—if I want to designate the reality here, for example, as +a, then I would have to designate the reality within the wire as −a. So we have here an absorbing of something instead of what is actually always seen as something flowing in. Essentially what we are dealing with is the fact that, if there is an electrical conductor there, it doesn't actually constitute something that fills up. Rather it constitutes a hollow space in the spiritual. And now that takes us to the nature of the will, which Dr. Stein only touched upon before, and which is also based on the fact that with nerves, for example, we are not dealing with something that fills up, but with hollow channels, hollow tubes, through which the spiritual is drawn on and through which the spiritual passes.

But, as I said, that would lead too far afield today, and all I have been able to do was to take up the task of showing to what degree, or rather, *how* I meant it when I said that these more recent phenomena are actually in line with the further development of Goethean color theory.

Notes

The notes are by the editors of the German edition (with English editions substituted for the works cited, when available) unless otherwise noted. Works of Rudolf Steiner from the *Complete Works (Gesamtausgabe = GA)* are cited by the bibliography number.

Translator's Introduction

1. Stephen Edelglass, Georg Maier, et al., *The Marriage of Sense and Thought* (Hudson, NY: Lindisfarne Books, 1997), p. 135.

First Lecture

1. *Following up on the words just read to us here:* At the beginning of the course Walter Johannes Stein read aloud the following quotation from Steiner:

> Naturally, I would not dream of trying to defend all the details of Goethean color theory. What I do want to uphold is the *principle*. But even there it cannot be my task to derive from his principle phenomena of color theory that were unknown in Goethe's time. If someday I should be so lucky as to have the leisure and means to write a color theory in the Goethean sense that is current with the latest achievements of modern natural science, that would be the only way to solve this problem. (From the introduction to *Goethes naturwissenschaftlichen Schriften* [Goethe's Natural Scientific Writings], edited and annotated by Rudolf Steiner 1884–97 in Kürschner's *Deutsche National-Litteratur,* in five volumes, GA 1a–e, [reprinted Dornach: Rudolf Steiner Verlag, 1975], vol. 3, p. XVII; p. 279 of the special edition of the complete introductions.)

Let youthfully striving thinkers and researchers, especially those who are not interested only in details, but who take the central question of our knowing head on, give some attention to my remarks and follow in droves to carry out more perfectly what I have been trying to carry out. (In the above cited introductions, vol. 1, p. LXXXIV; p. 120 of the special edition.)

In the future, chemists and physicists will come who will not teach chemistry and physics in the way they are taught today under the influence of the remaining Egyptian-Chaldean spirits, but who will teach: 'Matter is constructed in the way that Christ arranged it bit by bit.' We will look for Christ even in the laws of chemistry and physics. A spiritual chemistry, a spiritual physics is what will come in the future. (In *The Spiritual Guidance of the Individual and Humanity* [1911], GA 15 [Hudson, NY: Anthroposophic Press, 1992], p. 56.)

2. *We can continue what we have begun:* A second course on the natural sciences (known as the Warmth Course) took place from March 1 to March 14, 1920, then the course *Das Verhältnis der verschiedenen naturwissenschaftlichen Gebiete zur Astronomie* ("Relation of Diverse Branches of Natural Science to Astronomy") from January 1 to January 18, 1921, both in Stuttgart, GA 321 and 323. A guide to all the lectures on the natural sciences is found in *Bibliographische Übersicht,* vol. 1 of the indices to Rudolf Steiner's complete works.

3. *Lecture on Goethe's natural science:* Given on August 27, 1893. Printed under the title "Goethes Naturanschauung gemäß den neuesten Veröffentlichungen des Goethe-Archivs" ("Goethe's View of Nature, According to the Most Recent Publications of the Goethe Archive") in *Methodische Grundlagen der Anthroposophie 1884–1901* (Methodological Foundations of Anthroposophy), GA 30 (Dornach: Rudolf Steiner Verlag, 1961), p. 69. Cf. *Autobiography,* GA 28 (Hudson, NY: Anthroposophic Press, 1999), p. 220.

4. *Here I already falter:* Goethe's *Faust,* Part I, "The Study."

5. *Now we have grown used to recognizing the smallest thing … by its effect:* The paragraph beginning with these words exemplifies how the lecturer shapes many different things in a free-ranging account. In a packed summary the audience is given a picture of both the

unity of force and the atomistic approach to thinking. Force and the unit of force are not characterized in the usual way, but by means of the transferred impulse, in accordance with the atomistic approach (cf. also the beginning of the Second Lecture), which originally has the goal of composing phenomena out of indivisible units or quanta, thereby fulfilling the connection with the concept "unit." In line with the development that started at the beginning of the twentieth century, the exemplification of atomism leads from the realm of matter into that of "force." By bringing the quantum of force or of the impulse together with the ordinary unit, the portrayal sketches out a picture of the way of thinking in a few strokes. Cf. in this regard the methodological point of view at the beginning of the Seventh Lecture.

Second Lecture

1. *etheric body:* Cf. *Theosophy: An Introduction to the Spiritual Processes in Human Life and in the Cosmos* (1904), GA 9 (Hudson, NY: Anthroposophic Press, 1994). On the types of ether cf. *Cosmic Memory* (1904–08), GA 11 (Blauvelt, NY: Garber Communications, 1998).

2. Cf. *Theosophy* on the *etheric body* and the *astral body.* For an ophthalmologic perspective, the reader is also referred to the discussion on the relationship of the astral body to the etheric body in *Physiologisch-Therapeutisches auf Grundlage der Geisteswissenschaft,* GA 314 (Dornach: Rudolf Steiner Verlag, 1975), pp. 316, etc.

Third Lecture

1. *Waldorf School:* Founded in 1919 by the businessman Emil Molt for the children of the workers and office personnel of the Waldorf Astoria Cigarette Factory in Stuttgart. The school was established and directed by Rudolf Steiner.

2. *In Goethe you can read:* "Materialien zur Geschichte der Farbenlehre, Konfession des Verfassers," in *Goethes naturwissenschaftlichen Schriften* (cf. note 1 to First Lecture), reference to p. 15, vol. 5, p. 128.

3. The unorthodox figure is confirmed by drawings in the notebook entries made by Steiner at the time of the *Light Course*. Cf. *Geisteswissenschaftliche Impulse zur Entwickelung der Physik*, GA 320 (Dornach: Rudolf Steiner Verlag, 1987), p. 186. Interestingly, first there is the usual figure and then below it the new figure showing the lifting. The latter recalls the appearance of a rod that dips at an angle into a well.

Fourth Lecture

1. *A diversion would take place:* Cf. Figure 6a in the Sixth Lecture.

2. *Isaac Newton* (1643–1727). English physicist, mathematician, and astronomer.

3. *Christian Huygens* (1629–1695). Dutch physicist, mathematician, and astronomer.

4. *Thomas Young* (1773–1829). English natural scientist, Egyptologist.

5. *Augustin-Jean Fresnel* (1788–1827). French engineer and physicist.

6. *Francesco Maria Grimaldi* (1618–1663). Italian mathematician and physicist.

7. *Leonhard Euler* (1707–1783). Swiss mathematician, astronomer, and physicist.

8. *bluish green:* Cf. description of Figure 2c in the Second Lecture.

9. *That's what made Goethe lose his faith:* See note 2 to Third Lecture.

Fifth Lecture

1. *The experiment of Bunsen and Kirchhoff:* Gustav Robert Kirchhoff, "Über die Fraunhoferschen Linien," *Monatsbericht der Akademie der Wissenschaft zu Berlin,* October 1859; *Gesammelte Abhandlungen* (Leipzig: 1882).

2. *A shoemaker in Bologna:* Vincenzo Cascariolo; cf. the footnote in *Goethes naturwissenschaftliche Schriften* (cf. note 1 to First Lecture), vol. 5, p. 146.

Sixth Lecture

1. *Draw a shining circle here:* In the copy of a blackboard drawing the white line is reproduced in black in the figure, thereby reversing black and white.

2. *A brighter spot and a somewhat darker spot:* See note 1 above.

3. *People like Kirchhoff, for example:* Gustav Robert Kirchhoff (1824–87). German physicist. There are a few words about his scientific point of view in the preface to his work on mechanics, *Vorlesungen über mathematische Physik,* Vol. 1 (Leipzig: 1876). Cf. *Riddles of Philosophy,* GA 18 (Spring Valley, NY: Anthroposophic Press, 1973), p. 323. Ludwig Boltzmann writes about the reception and the effects of this point of view in *Gustav Robert Kirchhoff* (Leipzig: 1888).

4. *There have been people who said that's nonsense:* Heinrich Schramm, *Die allgemeine Bewegung der Materie als Grundursache aller Naturerscheinungen* (Vienna: 1872); cf. *Autobiography,* GA 28 (Hudson, NY: Anthroposophic Press, 1999), pp. 31–32.

5. *Hermann Helmholtz* (1821–1894). German physicist and physiologist. The thought mentioned in the lecture occupied all of his research in the last two years of his life, as in the treatise "Folgerungen aus Maxwells Theorie über die Bewegungen des reinen Äthers," 1893.

Seventh Lecture

1. *If you take a small tube and look through it ... then you will also see it as green:*

This experiment was repeatedly attempted, always with negative results, by V. C. Bennie, lecturer in physics at that time at Kings College of the University of London, after he had read the transcription of the course by Rudolf Steiner in 1921. Because of this, there were two evenings of experiments in Dornach at the end of September 1922. Rudolf Steiner had wished to be present. The other collaborators were Dr. Ernst Blümel, mathematician, Bennie, and Dr. Oskar Schmiedel, pharmacist and director of courses on Goethe's theory of color. On the first evening, Dr. W. J. Stein also participated. The two evenings did not lead to a confirmation of the experiment with the tube. Incidentally the result was reported differently by the participants. What is important here, however, does not seem to have been discussed at all on the two evenings, namely Rudolf Steiner's intention, as reported by Dr. Blümel, to prove the objectivity of the color in the shadow by photographic or chemical means in the Stuttgart research institute. However, nothing is known of such experiments— and certainly not with positive results—of the research institute at that time. Later, when the first edition of the course was to appear in the *Complete Works*, there were photographic experiments available with negative results: despite the advances in color photography since the time of Rudolf Steiner, the color in the photographs of the colored shadows was not stable. The whole picture did indeed show the shadow in the required color, but when cut out, it appeared gray. Today that is different. Stable colors result even without special procedures. The starting point of new experiments was a photograph that the professional photographer and elaborator of Goethe's color theory Hans-Georg Hetzel was able to make of an experiment with colored shadow in the Goethe-Color-Studio in Dornach. Besides the usual trinity of demanding color, colored shadow, and brightened color of the surrounding field, the photograph also showed a small technical gray scale. Despite the intense color of the shadow the latter appeared gray, on the same photograph!

Today there are series of photographs available of different kinds of colored shadows, which can be reproduced by Hans-Georg Hetzel, each series being photographed on the same film and supplemented for control purposes by interposed photographs of gray shadow. These are slide films. Each film is developed professionally by machine as one among many customer orders. Thus the different colors of a series are produced in one and the same developing process. Even the photographs were taken in a uniform way: in every case the lens was fitted with a transparency of the same color—the color that the color temperature meter indicated for photographing gray so that the gray really turns out gray. If this condition is not fulfilled, then a decision still must made: either all the colored shadows appear as gray, so the colors of the shadows could be subjective, or the shadows appear different from the gray, so a special effect is taking place in that space. That the latter is the case is shown by the special color process of the Polaroid camera, which gives the shadow a strongly green cast, unlike the gray. There cannot be any question of the colored shadows coming out like the gray ones. If it were only a matter of subjective and objective, it could be left at that. However, if we want to come as close as possible to the true colors, it is necessary, of course, for gray to turn out gray. If we describe the best of the resulting series, the gray is a beautiful mouse gray. The blue shadow appears gray with at most a hint of blue. The other shadows are more decidedly colored, all of them with a brownish cast, in comparison with which the color called for is revealed only as a nuance. Even green turns out decidedly different from gray, but in a shade that is difficult to evaluate and that is usually described as brownish. If enlarged in an automatic process and copied onto paper, the series shows blue and green the same, and in the rest the brown shade dominates to the extent that the other nuances disappear. It has already been indicated that the film type plays an important role. It is interesting to note, however, that the quality of lighting is also significant. Diffuse light (e.g., stage lights) provide better colors than harshly focused light. Individual photographs of colored shadows have been gotten with very beautiful, stable color. Their beauty is achieved, however, by means of special treatment of the individual photograph, so that they do not have the same value as evidence. Any photograph, however, that results from procedures that are also routinely employed for photographing ordinary colors can be regarded as evidence, since it shows that the photographic process that was developed for ordinary colors also reacts to colored shadows. Nothing more than this is being asserted here. For the whole question of colored shadows, cf. G. Ott and H. O. Proskauer, "Das Rätsel des farbigen Schattens" (Basel: 1979). A series of the photographs mentioned above is located in the archives of the Rudolf-Steiner-Nachlassverwaltung (Rudolf Steiner Estate Administration), Dornach. More details about the experiments are set out in *Beiträgen zur Rudolf Steiner Gesamtausgabe,* issue number 97, Michaelmas 1987.

2. *The truth be known when two are shown (Durch zweier Zeugen Mund wird alle Wahrheit kund): Faust,* Part I, "The Neighbor Woman's House."

3. *When I breathe in again, the cerebrospinal fluid is pushed upward:* The exact discussion of this was transcribed only in fragmentary form.

4. *But with what terminology:* The word "demonology" is in the transcription instead of "terminology."

5. *Where we turn into beings of the air . . . outer air:* The sentence, which was incomplete in the transcription, was completed by the words, "outer air," in line with the preceding description.

6. *Julius Robert Mayer* (1814–1878). German physician and physicist.

Eighth Lecture

1. *Leonardo da Vinci* (1452–1519). The great Italian artist, designer, and inventor of the Renaissance.

2. *Marin Mersenne* (1588–1648). French mathematician and music theoretician.

3. *With a pencil attached to it:* A vibrating tuning fork is passed along a plate covered with soot. The pencil attached to one of the tines draws the form of a wave in the soot.

4. *The expression we are accustomed to hearing from Goethe:* For example, "so the splendid man did not think that there was a difference between seeing and seeing, that the eyes of the spirit have to remain in a constant living union with the eyes of the body because otherwise we run the danger of seeing and yet overseeing" ("History of My Botanical Studies"). More in *Goethes naturwissenschaftliche Schriften* (cf. note 1 to the First Lecture), vol. 1, p. 107.

5. *E. A. Karl Stockmeyer* (1886–1963). Mathematician, philosopher, and teacher at the Freie Waldorfschule in Stuttgart.

6. *Robert Hamerling* (1830–1889). Poet and philosopher. The citation is from *Riddles of Philosophy*, GA 18 (Spring Valley, NY: Anthroposophic Press, 1973).

7. *Certain organs such as the pecten . . . or the falciform process:* The pecten is a fan-shaped vascular organ that extends into the vitreous

humor in the eye of nearly all birds and many reptiles; the falciform process, literally *sickle-shaped* process, is a similar blood-filled organ found in certain fishes. —*Trans.*

Ninth Lecture

1. *Luigi Galvani* (1737–1798). Italian physician and natural scientist.

2. *Alessandro Volta* (1745–1827). Italian physicist.

3. *Heinrich Hertz* (1857–1894). German physicist.

4. *William Crookes* (1832–1919). British physicist and chemist.

5. *Johann Wilhelm Hittorf* (1824–1914). German physicist.

6. *Philipp Lenard* (1862–1947). German physicist.

7. *Wilhelm Conrad Röntgen* (1845–1923). German physicist.

8. *The alpha rays at about one-tenth the speed of light:* Rutherford's first measurements (1902) yielded one-twelfth of the speed of light for radium; later lower values were found, approximately one-twentieth of the speed of light.

9. *Luciferic and ahrimanic influences:* Cf. *An Outline of Esoteric Science,* GA 13 (Hudson, NY: Anthroposophic Press, 1997), pp. 197–280.

10. *Something that radiates as an emanation of the substance radium:* Radium radiation. He is not speaking of radium lead here; however, he does a short time later in a medical lecture, in *Introducing Anthroposophical Medicine,* GA 312 (Hudson, NY: Anthroposophic Press, 1999), p. 172.

11. *What an English philosopher said:* A. J. Balfour in his speech at the British Association, 1904. Cf. *Lucifer Gnosis 1903–1908. Gesammelte Aufsätze,* GA 34 (Dornach: Rudolf Steiner Verlag, 1987), p. 467.

Tenth Lecture

1. *Then these substances … emit yet a fourth substance, which is the element itself:* Cf. note 10 to the Ninth Lecture.

2. *Nikolai Ivanovich Lobachevsky* (1793–1856). Russian mathematician noted for his work in non-Euclidean geometry.

3. *When we ascend to intuition:* Cf. *How to Know Higher Worlds* (1904–1905), GA 10 (Hudson, NY: Anthroposophic Press, 1994).

4. *Luciferic/light-bearing and ahrimanic/electromagnetic:* Cf. note no. 9 to the Ninth Lecture.

5. *Which I hope to be able to expand on:* Cf. note no. 2 to the First Lecture.

6. *In the pedagogical course when instruction began at the Waldorf School: The Foundations of Human Experience* (a cycle of fourteen lectures, held in Stuttgart from August 21 to September 5, 1919, on the occasion of the founding of the Free Waldorf School), GA 293 (Hudson, NY: Anthroposophic Press, 1996).

7. *Hermann Grimm* (1828–1901). German art historian. The quotation can be found in *Goethe* (Berlin: 1877), vol. 2, Lecture 23, p. 171.

Discussion Statement

1. *Discussion Statement from Rudolf Steiner on August 8, 1921:* The statement refers to a report by Dr. Rabel on "Conflicting Effects of Light." The text, which has not previously appeared in print, exists in shorthand form with considerable gaps in the transcription that could not always be filled out.

2. *One of her treatises:* Gabriele Rabel, "Farbenantagonismus oder die chemische und electrische Polarität des Spektrums." Offprint from the *Zeitschrift für wissenschaftliche Photographie,* vol. 19 (1919).

3. *A physicist with whom I had a conversation about Goethean color theory:* Salomon Kalischer, the editor of Goethe's color theory in the *Sophienausgabe;* cf. *Autobiography,* GA 28 (Hudson, NY: Anthroposophic Press, 1999), p. 221.

4. *Lecture . . . about Goethe's view of nature:* See note 3 to the First Lecture.

5. *Oskar Schmiedel* (1887–1959). Chemist, for many years director of Weleda Inc., in Arlesheim and Schwäbisch-Gmünd.

6. *Opposition between red nature and blue nature:* The reading of this and the following sentences is very uncertain. "Nature" is fairly clear. "Entity" can be read perhaps. There are gaps in the accompanying words, and the text is difficult to decipher in places.

7. *Eugen Dreher* (1841–1900). Cf. his "Beiträge zu unserer modernen Atom- und Molekular-Theorie auf kritischer Grundlage" (Halle: 1882), p. 67, and the extensive footnote in *Goethes naturwissenschaftlichen Schriften* (cf. Note 1 to First Lecture), vol. 5, p. 147.

8. *Is the beginning of the paths to higher knowledge:* Cf. *How to Know Higher Worlds* (1904–1905), GA 10 (Hudson, NY: Anthroposophic Press, 1994).

9. *There can be no satisfactory theory of knowledge:* Cf., in addition to Rudolf Steiner's epistemological writings, "The Psychological Foundations of Anthroposophy: Its Standpoint in Relation to the Theory of Knowledge," lecture at the International Philosophical Conference in Bologna (1911), in *Philosophie und Anthroposophie: Gesammelte Aufsätze 1904–1918,* GA 35; published in English in the collection *Esoteric Development: Selected Lectures and Writings from the Works of Rudolf Steiner* (Spring Valley, NY: Anthroposophic Press, 1982).

10. *Louis Blanc* (1811–1882). French writer.

11. *Walter Johannes Stein* (1891–1957). Austrian writer and lecturer; teacher at theWaldorf School in Stuttgart.

Index

THE FOUNDATIONS
OF WALDORF EDUCATION

THE FIRST FREE WALDORF SCHOOL opened its doors in Stuttgart, Germany, in September 1919, under the auspices of Emil Molt, director of the Waldorf Astoria Cigarette Company and a student of Rudolf Steiner's spiritual science and particularly of Steiner's call for social renewal.

It was only the previous year—amid the social chaos following the end of World War I—that Emil Molt, responding to Steiner's prognosis that truly human change would not be possible unless a sufficient number of people received an education that developed the whole human being, decided to create a school for his workers' children. Conversations with the minister of education and with Rudolf Steiner, in early 1919, then led rapidly to the forming of the first school.

Since that time, more than six hundred schools have opened around the globe—from Italy, France, Portugal, Spain, Holland, Belgium, Britain, Norway, Finland, and Sweden to Russia, Georgia, Poland, Hungary, Romania, Israel, South Africa, Australia, Brazil, Chile, Peru, Argentina, Japan, and others—making the Waldorf school movement the largest independent school movement in the world. The United States, Canada, and Mexico alone now have more than 120 schools.

Although each Waldorf school is independent, and although there is a healthy oral tradition going back to the first Waldorf teachers and to Steiner himself, as well as a growing body of secondary literature, the true foundations of the Waldorf method and spirit remain the many lectures that Rudolf Steiner gave on the subject. For five years (1919–1924), Rudolf Steiner, while simultaneously working on many other fronts, tirelessly dedicated himself to the dissemination of the idea of Waldorf education. He gave manifold lectures to teachers, parents, the general public, and even the children themselves. New schools were founded. The movement grew.

· While many of Steiner's foundational lectures have been translated and published in the past, some have never appeared in English, and many have been virtually unobtainable for years. To remedy this situation and to establish a coherent basis for Waldorf education, Anthroposophic Press has decided to publish the complete series of Steiner lectures and writings on education in a uniform series. This series will thus constitute an authoritative foundation for work in educational renewal, for Waldorf teachers, parents, and educators generally.

RUDOLF STEINER'S LECTURES
AND WRITINGS ON EDUCATION

I. *Allgemeine Menschenkunde als Grundlage der Pädagogik: Pädagogischer Grundkurs,* 14 Lectures, Stuttgart, 1919 (GA 293). Previously *Study of Man. The Foundations of Human Experience* (Anthroposophic Press, 1996).

II. *Erziehungskunst Methodische-Didaktisches,* 14 Lectures, Stuttgart, 1919 (GA 294). *Practical Advice to Teachers* (Rudolf Steiner Press, 1988).

III. *Erziehungskunst,* 15 Discussions, Stuttgart, 1919 (GA 295). *Discussions with Teachers* (Anthroposophic Press, 1997).

IV. *Die Erziehungsfrage als soziale Frage,* 6 Lectures, Dornach, 1919 (GA 296). *Education as a Force for Social Change* (previously *Education as a Social Problem*; Anthroposophic Press, 1997).

V. *Die Waldorf Schule und ihr Geist,* 6 Lectures, Stuttgart and Basel, 1919 (GA 297). *The Spirit of the Waldorf School* (Anthroposophic Press, 1995).

VI. *Rudolf Steiner in der Waldorfschule, Vorträge und Ansprachen,* Stuttgart, 1919–1924 (GA 298). *Rudolf Steiner in the Waldorf School: Lectures and Conversations* (Anthroposophic Press, 1996).

VII. *Geisteswissenschaftliche Sprachbetrachtungen,* 6 Lectures, Stuttgart, 1919 (GA 299). *The Genius of Language* (Anthroposophic Press, 1995).

VIII. *Konferenzen mit den Lehrern der Freien Waldorfschule 1919–1924,* 3 Volumes (GA 300a–c). *Faculty Meetings with Rudolf Steiner,* 2 Volumes (Anthroposophic Press, 1998).

IX. *Die Erneuerung der pädagogisch-didaktischen Kunst durch Geisteswissenschaft,* 14 Lectures, Basel, 1920 (GA 301). *The Renewal of Education* (Anthroposophic Press, 2001).

X. *Menschenerkenntnis und Unterrichtsgestaltung,* 8 Lectures, Stuttgart, 1921 (GA 302). Previously *The Supplementary Course—Upper School* and *Waldorf Education for Adolescence. Education for Adolescents* (Anthroposophic Press, 1996).

XI. *Erziehung und Unterricht aus Menschenerkenntnis,* 9 Lectures, Stuttgart, 1920, 1922, 1923 (GA 302a). The first four lectures available as *Balance in Teaching* (Mercury Press, 1982); last three lectures as *Deeper Insights into Education* (Anthroposophic Press, 1988).

XII. *Die gesunde Entwicklung des Menschenwesens,* 16 Lectures, Dornach, 1921–22 (GA 303). *Soul Economy and Waldorf Education* (Anthroposophic Press, 1986).

XIII. *Erziehungs- und Unterrichtsmethoden auf anthroposophischer Grundlage,* 9 Public Lectures, various cities, 1921–22 (GA 304). *Waldorf Education and Anthroposophy 1* (Anthroposophic Press, 1995).

XIV. *Anthroposophische Menschenkunde und Pädagogik,* 9 Public Lectures, various cities, 1923–24 (GA 304a). *Waldorf Education and Anthroposophy 2* (Anthroposophic Press, 1996).

XV. *Die geistig-seelischen Grundkräfte der Erziehungskunst,* 12 Lectures, 1 Special Lecture, Oxford, 1922 (GA 305). *The Spiritual Ground of Education* (Garber Publications, 1989).

XVI. *Die pädagogische Praxis vom Gesichtspunkte geisteswissenschaftlicher Menschenerkenntnis,* 8 Lectures, Dornach, 1923 (GA 306). *The Child's Changing Consciousness as the Basis of Pedagogical Practice* (Anthroposophic Press, 1996).

XVII. *Gegenwärtiges Geistesleben und Erziehung,* 4 Lectures, Ilkeley, 1923 (GA 307). *A Modern Art of Education* (Rudolf Steiner Press, 1981) and *Education and Modern Spiritual Life* (Garber Publications, 1989).

XVIII. *Die Methodik des Lehrens und die Lebensbedingungen des Erziehens,* 5 Lectures, Stuttgart, 1924 (GA 308). *The Essentials of Education* (Anthroposophic Press, 1997).

XIX. *Anthroposophische Pädagogik und ihre Voraussetzungen,* 5 Lectures, Bern, 1924 (GA 309). *The Roots of Education* (Anthroposophic Press, 1997).

XX. *Der pädagogische Wert der Menschenerkenntnis und der Kulturwert der Pädagogik,* 10 Public Lectures, Arnheim, 1924 (GA 310). *Human Values in Education* (Rudolf Steiner Press, 1971).

XXI. *Die Kunst des Erziehens aus dem Erfassen der Menschenwesenheit,* 7 Lectures, Torquay, 1924 (GA 311). *The Kingdom of Childhood* (Anthroposophic Press, 1995).

XXII. *Geisteswissenschaftliche Impulse zur Entwickelung der Physik. Erster naturwissenschaftliche Kurs: Licht, Farbe, Ton—Masse, Elektrizität, Magnetismus,* 10 Lectures, Stuttgart, 1919–20 (GA 320). *The Light Course* (Anthroposophic Press, 2001).

XXIII. *Geisteswissenschaftliche Impulse zur Entwickelung der Physik. Zweiter naturwissenschaftliche Kurs: die Wärme auf der Grenze positiver und negativer Materialität,* 14 Lectures, Stuttgart, 1920 (GA 321). *The Warmth Course* (Mercury Press, 1988).

XXIV. *Das Verhältnis der verschiedenen naturwissenschaftlichen Gebiete zur Astronomie. Dritter naturwissenschaftliche Kurs: Himmelskunde in Beziehung zum Menschen und zur Menschenkunde,* 18 Lectures, Stuttgart, 1921 (GA 323). Available in typescript only as "The Relation of the Diverse Branches of Natural Science to Astronomy."

XXV. *The Education of the Child and Early Lectures on Education* (a collection; Anthroposophic Press, 1996).

XXVI. Miscellaneous.

DURING THE LAST TWO DECADES of the nineteenth century the Austrian-born Rudolf Steiner (1861–1925) became a respected and well-published scientific, literary, and philosophical scholar, particularly known for his work on Goethe's scientific writings. After the turn of the century he began to develop his earlier philosophical principles into an approach to methodical research of psychological and spiritual phenomena.

His multifaceted genius has led to innovative and holistic approaches in medicine, science, education (Waldorf schools), special education, philosophy, religion, agriculture (biodynamic farming), architecture, drama, movement, speech, and other fields. In 1924 he founded the General Anthroposophical Society, which today has branches throughout the world.